TRANSLATIONS SERIES IN MATHEMATICS AND ENGINEERING

TRANSLATION SERIES IN MATHEMATICS AND ENGINEERING

M.I. Yadrenko
SPECTRAL THEORY OF RANDOM FIELDS

1983, viii + 259 pp.
ISBN 0-911575-00-6 Optimization Software, Inc.
ISBN 0-387-90823-4 Springer-Verlag New York Berlin Heidelberg Tokyo
ISBN 3-540-90823-4 Springer-Verlag Berlin Heidelberg New York Tokyo

G.I. Marchuk
MATHEMATICAL MODELS IN IMMUNOLOGY

1983, xxv + 353 pp.
ISBN 0-911575-01-4 Optimization Software, Inc.
ISBN 0-387-90901-X Springer-Verlag New York Berlin Heidelberg Tokyo
ISBN 3-540-90901-X Springer-Verlag Berlin Heidelberg New York Tokyo

A.A. Borovkov, Ed.
ADVANCES IN PROBABILITY THEORY: LIMIT THEOREMS AND RELATED PROBLEMS

1984, xiv + 378 pp.
ISBN 0-911575-03-0 Optimization Software, Inc.
ISBN 0-387-90945-1 Springer-Verlag New York Berlin Heidelberg Tokyo
ISBN 3-540-90945-1 Springer-Verlag Berlin Heidelberg New York Tokyo

V.A. Dubovitskij
THE ULAM PROBLEM OF OPTIMAL MOTION OF LINE SEGMENTS

1985, xiv + 114 pp.
ISBN 0-911575-04-9 Optimization Software, Inc.
ISBN 0-387-90946-X Springer-Verlag New York Berlin Heidelberg Tokyo
ISBN 3-540-90946-X Springer-Verlag Berlin Heidelberg New York Tokyo

Yu. G. Evtushenko
NUMERICAL OPTIMIZATION TECHNIQUES

1985, approx. 450 pp.
ISBN 0-911575-07-3 Optimization Software, Inc.
ISBN 0-387-90949-4 Springer-Verlag New York Berlin Heidelberg Tokyo
ISBN 3-540-90949-4 Springer-Verlag Berlin Heidelberg New York Tokyo

V.A. DUBOVITSKIJ

THE ULAM PROBLEM OF OPTIMAL MOTION OF LINE SEGMENTS

OPTIMIZATION SOFTWARE, INC.
PUBLICATIONS DIVISION, NEW YORK

Author
V.A. DUBOVITSKIJ
OIKhF AN SSSR
p/o Chernogolovka
Noginskij rajon
Moskovskaya oblast'
USSR 142432

Series Editor
A.V. BALAKRISHNAN
School of Engineering
University of California
Los Angeles
California 90024
USA

Translated by JOHN T. ELLIS

Library of Congress Cataloging in Publication Data

Dubovitskij, V.A.
The Ulam problem of optimal motion of line segments.

(Translation series in mathematics and engineering)
Translated from Russian.
Bibliography: p.
Includes index.
1. Mathematical optimization. 2. Control theory.
3. Maximum principles (Mathematics). 4. Differential equations--Numerical solutions. I. Title. II. Series.
QA402.5.D8155 1985 519 84-19035
ISBN 0-911575-04-9

Exclusively authorized English translation of the original Russian edition of *Zadacha Ulama ob optimal'nom sovmeshchenii otrezkov*, published in 1981 by the Institute of Chemical Physics, the USSR Academy of Sciences, Chernogolovka, Moscow.

Worldwide Distribution Rights by Springer-Verlag New York, Inc., 175 Fifth Avenue, New York, NY 10010, USA and Springer-Verlag Berlin Heidelberg New York Tokyo, Heidelberg Platz 3, Berlin-Wilmersdorf-33, West Germany.

ISBN 0-911575-04-9 Optimization Software, Inc.
ISBN 0-387-90946-X Springer-Verlag New York Berlin Heidelberg Tokyo
ISBN 3-540-90946-X Springer-Verlag Berlin Heidelberg New York Tokyo

Cover by VLADO SHILINIS

ABOUT THE AUTHOR:

VLADIMIR ABRAMOVICH DUBOVITSKIJ is with the Institute of Chemical Physics of the USSR Academy of Sciences. He is a 1976 graduate of the Moscow Railroad Engineering Institute. He defended his Candidate dissertation entitled "Necessary and Sufficient Conditions for the Minimum in Optimal Control Problems with Singular and Sliding Regimes" at the Moscow State University in 1983.

FOREWORD

In Variational Theory and in Optimal Control Theory in particular, one rarely encounters significant problems whose solutions can be put in closed form when non-linearities are present. In this remarkable book, Dr. Dubovitskij has succeeded in solving in closed form, using the Dubovitsky-Milyutin theory, a generalization of a problem of Stanislaw Ulam, which stated succintly is: Among all continuous motions of an oriented line segment S in E^n from one position to another, which preserves its length and for which the endpoints of S lie on prescribed surfaces, find one for which the sum of the lengths of the paths swept by its endpoints is minimal. It should be noted that, while this problem formulates as a "bounded state" optimal control problem, usually additional hypotheses are imposed, which are not applicable here.

M.R. Hestenes
Professor Emeritus of Mathematics

UCLA

In the preparation of the English edition of his work ZADACHA ULAMA OB OPTIMAL'NOM SOVMESHCHENII OTREZKOV, the author has taken advantage of the opportunity for extensive revision as well as inclusion of additional material not in the original Russian publication. In particular, a whole new chapter, Chapter 5, has been added. An Appendix on the Maximum Principle has been included to make the work more self-contained. There are also ten new illustrations. These changes should serve to enhance the value of the work to the English-speaking audience.

EDITOR

CONTENTS

PREFACE

In this book we study the following problem: among the continuous motions of an oriented segment in E^n from one position to another, which preserve its length and for which the endpoints of the segment lie on prescribed surfaces, find the one for which the sum of the lengths of the paths swept out by the endpoints is minimal.

This is a generalization of a problem posed by S. Ulam in *A Collection of Mathematical Problems* [1] (in the original statement n was 2 and there were no conditions on the endpoints).

In recent years, problems like this have drawn the attention of many mathematicians, due likely to the simple geometric formulation on the one hand, and to the absence of solution techniques on the other. Mathematically, these problems reduce to an optimal control problem with nonsmooth constraints on state and control, to which neither the methods of classical Calculus of Variations nor Pontryagin's Maximum Principle is applicable.

Under various assumptions on the class of extremals (usually their piecewise smoothness) this problem has been solved by the following authors: Rvachev solved Ulam's problem for E^n [2], and for the Lobachevsky plane and the sphere [3]; Gurevich [4], Penrose [5], and Goldberg [6] solved Ulam's problem for E^2.

The possibility of solving Ulam's problem without any a priori assumptions on the nature of the extremals arose after the development of the integral maximum principle technique by A.Ya. Dubovitskij and A.A. Milyutin [7, 8]. Applying this principle, A.Ya. Dubovitskij [9] solved Ulam's problem for E^n without any a priori assumptions whatever on the nature of the extremals.

In this work, using the integral maximum principle, we investigate the general problem of motion of segments as stated. By a general variational analysis (analysis of a system of conditions resulting from the maximum principle), it is proved that the notion of extremality in Ulam's problem is equivalent to a system of differential equations, which is analogous to an equation for geodesics in classical Differential Geometry.

In a number of specific versions of Ulam's problem, these differential equations are amenable to complete analytic study. We have thus been able to get a full description of extremals in the following problems: optimal motion in E^n, on the n-sphere, on the two-dimensional circular cylinder, and also on the anisotropic plane (i.e., when different Euclidean metrics measure the length of the segment and the paths swept out by the endpoints).

Chapter 5 is devoted to the problem of synthesizing optimal motions of segments in the space E^n. In the case of a plane, the synthesis is reduced to an atlas of optimal motions, which makes it possible to find the required optimal motion by ruler-and-compass methods.

The author is grateful to A.A. Milyutin for posing the problem.

CHAPTER 1

GENERAL VARIATIONAL STUDY OF THE ULAM PROBLEM

We put our problem in the standard form of an optimal control problem.

We introduce the basic notation:

• 1. The functions $x(t)$, $y(t)$ describe the motion of the endpoints of the segment;

• 2. the surfaces M_1, M_2 on which the endpoints x, y of the segment remain are specified by vector equations $g_1(x) = 0$, $g_2(y) = 0$, respectively. Here g_1, g_2 are twice continuously differentiable functions mapping the initial space E^n into E^{k_1}, E^{k_2}, and at points of M_1, M_2 the tangent mappings g_1', g_2' are epimorphisms;

• 3. we will denote by $V_1(x)$, $V_2(y)$ the tangent subspaces to M_1, M_2 at x, y, respectively;

• 4. x_0^*, y_0^* and x_1^*, y_1^* denote the initial and final position of the endpoints of the segment;

• 5. for a subspace $L \subset E^n$ we will let P_L denote the operator of orthogonal projection onto L.

We will call the position of the segment $[x,y]$ with endpoints on M_1, M_2 a "general" position if $P_{V_1(x)}(x-y) \neq 0$, $P_{V_2(y)}(x-y) \neq 0$. We will only study the optimal motions of segments for which all assumable positions are general ones.

As a parameter describing the motion of the segment we adopt the sum of the lengths of the paths swept out by its endpoints. For the motion of a segment so parametrized the optimality criterion is that the trajectory x, y, $u = \dot{x}$, $v = \dot{y}$ is the solution of the following optimal control problem:

$$I = t_1 \to \min \quad \text{for } t_0 = 0,\ x_0 = x_0^*,\ y_0 = y_0^*,$$

$$P_L(x_1 - x_1^*,\ y_1 - y_1^*) = 0,$$

$$|x_1 - x_1^*| + |y_1 - y_1^*| \leq \varepsilon,$$

$$\dot{x} = u, \quad \dot{y} = v, \quad |u| + |v| \leq 1,$$

$$(x-y,\ u-v) = 0,$$

$$g_1'(x)u = 0, \quad g_2'(y)v = 0. \tag{1}$$

Here $\varepsilon > 0$ is sufficiently small,

$$L = \{(u,v) \mid (x_1^* - y_1^*,\ u-v) = 0,\ g_1'(x_1^*)u = 0,\ g_2'(y_1^*)v = 0\}.$$

The only thing that may not be completely obvious here is the uniqueness of the final position of the segment. Let us prove this.

Consider the function

$$G(x, y) = (|x - y|^2 - 1)/2,\ g_1(x),\ g_2(x))$$

(we agree to assume the segment has length one). Obviously, $L = \mathrm{Ker}\, G'(x_1^*,y_1^*)$ and from the general position of $[x_1^*,y_1^*]$ it follows that $G'(x_1^*,y_1^*)$ is an epimorphism. Since $|x_0^*-y_0^*|^2 = 1$, $g_1(x_0^*) = 0$, $g_2(y_0^*) = 0$, then from $(x-y,u-v) = 0$, $g_1'(x)u = 0$, $g_2'(y)v = 0$ it follows that $|x_1-y_1| = 1$, $g_1(x_1) = 0$, $g_2(y_1) = 0$, i.e., $G(x_1,y_1) = G(x_1^*,y_1^*) = 0$. By the differentiability of G,

$$G(x_1,y_1) = G(x_1^*,y_1^*) + G'(x_1^*,y_1^*)(x_1-x_1^*,\ y_1-y_1^*) + o(x_1-x_1^*,\ y_1-y_1^*) .$$

Whence

$$G'(x_1^*,y_1^*)(x_1-x_1^*,\ y_1-y_1^*) = o(x_1-x_1^*,\ y_1-y_1^*) .$$

But $G'|L^\perp$ is nondegenerate and $P_L(x_1-x_1^*,\ y_1-y_1^*) = 0$. Hence

$$(x_1-x_1^*,\ y_1-y_1^*) = G'|_{L^\perp}^{-1}\, o(x_1-x_1^*,\ y_1-y_1^*) .$$

Take ε so small that

$$\frac{|o(x_1-x_1^*,\ y_1-y_1^*)|}{|x_1-x_1^*| + |y_1-y_1^*|} \le \frac{1}{2\|G'|_{L^\perp}^{-1}\|} .$$

Then

$$|x_1-x_1^*| + |y_1-y_1^*| \le \frac{|x_1-x_1^*| + |y_1-y_1^*|}{2} .$$

Whence $x_1 = x_1^*$, $y_1 = y_1^*$, Q.E.D.

§ 1. REGULARITY OF THE LUMPED CONSTRAINTS

Let us show that the lumped constraints of problem (1): $(x-y,\ u-v) = 0$, $g_1'(x)u = 0$, $g_2'(y)v = 0$, $|u| + |v| \le 1$ are regular in u, v, i.e., for any admissible point x,y,u,v there exists a direction $\bar{u},\bar{v}$ satisfying the equality constraints and such that for small $\varepsilon > 0$, one has $|u+\varepsilon\bar{u}| + |v+\varepsilon\bar{v}| \leq 1$.

Obviously, if $|u| + |v| > 0$, one can take $\bar{u} = -u$, $\bar{v} = -v$. But if $u = 0$, $v = 0$, then $\bar{u},\bar{v}$ can be any nonzero pair of vectors such that $g_1'(x)\bar{u} = 0$, $g_2'(y)v = 0$.

§ 2. THE MAXIMUM PRINCIPLE

Since problem (1) has regular lumped constraints, as a necessary condition of minimality of the trajectory x,y,u,v one has the maximum principle for regular problems. According to the maximum principle, there exist conjugate multipliers ψ_1, ψ_2, ψ_t, λ, m, m_1, m_2, α_I, α, c_{t_0}, c_{x_0}, c_{y_0}, c_{x_1}, c_{y_1}, where

- a. ψ_1, ψ_2, ψ_t are Lipschitz functions on $[0,t_1]$;
- b. λ, m, m_1, m_2 are bounded measurable functions on $[0,t_1]$;
- c. ψ_t, λ, m are scalar functions, and ψ_1, ψ_2, m_1, m_2 are vector functions with values in E^n, E^n, E^{k_1}, E^{k_2}, respectively;
- d. α_I, α, $c_{t_0} \in R$, c_{x_0}, $c_{y_0} \in E^n$, $(c_{x_1},c_{y_1}) \in L$, for

which the following conditions are satisfied:

- 1. α_I, α, $\lambda \geq 0$,

$$\alpha(|x_1-x_1^*| + |y_1-y_1^*| - \varepsilon) = 0 ,$$

$$\lambda(|u| + |v| - 1) = 0 ; \tag{3}$$

$$\alpha_I + \alpha + |c_{t_0}| + |c_{x_0}| + |c_{y_0}| + |c_{x_1}| + |c_{y_1}| > 0 ;$$

- 2. the local maximum principle $H'_{u,v} = 0$;
- 3. the conjugate system

$$\frac{d}{dt}(\psi_1, \psi_2, \psi_t) = -H'_{x,y,t} ,$$

$$(\psi_1, \psi_2, \psi_t)\Big|_{t=0} = \ell'_{x_0, y_0, t_0} ,$$

$$(\psi_1, \psi_2, \psi_t)\Big|_{t=t_1} = -\ell'_{x_1, y_1, t_1} ;$$

• 4. the maximum principle proper

$$\max((\psi_1, u) + (\psi_2, v) + \psi_t) = 0 ,$$

$$|u| + |v| \leq 1 , \qquad (x-y,\ u-v) = 0 ,$$

$$g_1'u = 0 , \qquad g_2'v = 0 ,$$

and is attained a.e. on $u(t)$, $v(t)$.

Here H is the Hamiltonian function and ℓ is the terminal Lagrangian defined by the formulas:

$$H = (\psi_1, u) + (\psi_2, v) + \psi_t - \lambda(|u| + |v| - 1) - m(x-y, u-v)$$
$$- (m_1, g_1'(x)u) - (m_2, g_2'(y)v) ;$$

$$\ell = \alpha_I t_1 + \alpha(|x_1 - x_1^*| + |y_1 - y_1^*| - \varepsilon) + c_{t_0} t_0$$
$$+ \sum_{i=0}^{1}((c_{x_i},\ x_i - x_i^*) + (c_{y_i},\ y_i - y_i^*)) . \tag{4}$$

Note that under the conditions of the maximum principle (3) all equalities involving u, v hold a.e.

By a solution of problem (1) we will mean a complete description of the trajectories x,y,u,v for which one can choose conjugate multipliers so that the conditions (3) hold. Such trajectories are usually called extremals.

Since $x_1 = x_1^*$, $y_1 = y_1^*$, it follows from 3.1 that $\alpha = 0$. Noting this and using (4) we write the conditions (3) explicitly:

- 1. $\alpha_I, \lambda \geq 0$, $\lambda(|u| + |v| - 1) = 0$,

$$\alpha_I + |c_{t_0}| + |c_{x_0}| + |c_{y_0}| + |c_{x_1}| + |c_{y_1}| > 0 ;$$

- 2. $\psi_1 = \lambda n_u + mr + g_1'^* m_1$, $\psi_2 = \lambda n_v - mr + g_2'^* m_2$,

where $r = x - y$; (5)

- 3. $\dot{\psi}_1 = m(u-v) + g_1''^* u, m_1$, $\dot{\psi}_2 = m(v-u) + g_2''^* v, m_2$;

$$\dot{\psi}_t = 0 , \quad (\psi_1, \psi_2, \psi_t)\big|_{t=0} = (c_{x_0}, c_{y_0}, c_{t_0}) ,$$

$$(\psi_1, \psi_2, \psi_t)\big|_{t=t_1} = (-c_{x_1}, -c_{y_1}, -\alpha_I) ;$$

- 4. $\max\,((\psi_1, u) + (\psi_2, v) + \psi_t) = 0$;

$$|u| + |v| \leq 1 , \quad (r, u-v) = 0 ,$$

$$g_1' u = 0 , \quad g_2' v = 0 ,$$

and is attained a.e. on $u(t), v(t)$.

Let us analyze (5). From 5.3 we conclude that $\psi_t \equiv c_{t_0} = -\alpha_I$. Whence 5.4 yields $(\psi_1, u) + (\psi_2, v) = \alpha_I$. On the other hand, relations 5.2 give us

$$(\psi_1, u) + (\psi_2, v) = \lambda((n_u, u) + (n_v, v)) = \lambda(|u| + |v|) .$$

Now from the property $\lambda(|u| + |v| - 1) = 0$ we conclude that $\lambda \equiv \alpha_I$. Let us show that $\alpha_I > 0$. Suppose not. Then since $\alpha_I \geq 0$, $\alpha_I = 0$ and so $\lambda \equiv 0$. Now the local maximum principle 5.2 takes the form

$$\psi_1 = mr + g_1'^* m_1 \quad , \qquad \psi_2 = -mr + g_2'^* m_2$$

or

$$G'^*(m, m_1, m_2) = (\psi_1, \psi_2) \quad ,$$

whence

$$(m, m_1, m_2) = (G' \circ G'^*)^{-1} \circ G'(\psi_1, \psi_2) \quad .$$

Thus varying m, m_1, m_2 by measure 0, we get Lipschitz functions. Now substituting 5.2 into the conjugate system 5.3 we get

$$\dot{m}r + g_1'^* \dot{m}_1 = 0 \quad , \qquad -\dot{m}r + g_2'^* \dot{m}_2 = 0$$

or

$$G'^*(\dot{m}, \dot{m}_1, \dot{m}_2) = 0 \quad .$$

The latter, since $G' \circ G'^*$ is nondegenerate, gives $\dot{m}, \dot{m}_1, \dot{m}_2 = 0$. Thus m, m_1, m_2 = const. According to the right boundary conditions of the conjugate system:

$$\psi_1(t_1) = mr_1 + g_1'^* m_1 = -c_{x_1} \quad ,$$

$$\psi_2(t_1) = -mr_1 + g_2'^* m_2 = -c_{y_1}$$

or

$$G'^*(m, m_1, m_2) = -(c_{x_1}, c_{y_1}) \quad .$$

But

$$\operatorname{Im} G'^* = L^\perp \quad , \qquad (c_{x_1}, c_{y_1}) \in L \quad .$$

Hence $c_{x_1}, c_{y_1} = 0$ and since $G' \circ G'^*$ is nondegenerate, then $m, m_1, m_2 = 0$ and along with them $c_{x_0}, c_{y_0} = 0$. Thus all the conjugate multipliers are zero, contracting 5.1. The contradiction shows that $\alpha_I > 0$. Let us divide all conjugate multipliers by α_I. We then get a new set of conjugate multipliers, but $\lambda \equiv \alpha_I = 1$.

Thus the stationary conditions (5) are equivalent to the following:

$$\begin{aligned}
&\bullet\ 1. \quad |u| + |v| \equiv 1 ;\\
&\bullet\ 2. \quad \psi_1 = n_u + mr + g_1'*m_1 , \qquad \psi_2 = n_v - mr + g_2'*m_2 ;\\
&\bullet\ 3. \quad \dot\psi_1 = m(u-v) + g_1''*u,\ m_1 ,\\
&\qquad\quad \dot\psi_2 = m(v-u) + g_2''*v,\ m_2 , \qquad (\psi_1,\psi_2)\Big|_{t_1} \in L ;\\
&\bullet\ 4. \quad (\psi_1,u) + (\psi_2,v) \equiv 1 .
\end{aligned} \tag{6}$$

We have replaced condition 5.4 by the weaker condition 6.4 for convenience; we do not thereby lose the equivalence of (5) and (6) since 5.4 follows from 6.2.

§ 3. ANALYSIS OF THE LOCAL MAXIMUM PRINCIPLE

The local maximum principle together with other simple conditions leads us to stationary conditions considerably more tractable than (6). First of all, let us show that changing u, v, n_u, n_v, m, m_1, m_2 on a set of measure 0 we can attain that everywhere on $[0,t_1]$ the following relations are satisfied:

$$\begin{aligned}
&\psi_1 = n_u + mr + g_1'*m_1 ;\\
&\psi_2 = n_v - mr + g_2'*m_2 ;\\
&|u| + |v| = 1 , \qquad (r,\ u-v) = 0 ;\\
&g_1'u = 0 , \qquad g_2'v = 0 ;\\
&(\psi_1,u) + (\psi_2,v) = 1 .
\end{aligned} \tag{7}$$

Indeed, let E be a set of complete measure on $[0,t_1]$,

where these relations hold. The first two equations can be rewritten as

$$G'^*(m, m_1, m_2) = (\psi_1 - n_u, \psi_2 - n_v) .$$

Whence

$$(m, m_1, m_2) = (G' \circ G'^*)^{-1} \circ G'(\psi_1 - n_u, \psi_2 - n_v) .$$

Since

$$|u|, |v|, |n_u|, |n_v| \leq 1 ,$$

the set

$$X = \left\{ (t, u, v, n_u, n_v, m, m_1, m_2)\Big|_t \;\Big|\; t \in E \right\}$$

is bounded. Relying on this, we make the following construction: for any $t^* \in [0, t_1] \setminus E$,

$$u, v, n_u, n_v, m, m_1, m_2\Big|_{t^*} = u^*, v^*, n_u^*, n_v^*, m^*, m_1^*, m_2^* ,$$

where $(t^*, u^*, v^*, n_u^*, n_v^*, m^*, m_1^*, m_2^*)$ is a limit point of X. Thus, by changing u, v, n_u, n_v, m, m_1, m_2 by measure 0 we get (7) to hold everywhere on $[0, t_1]$. Here obviously n_u, n_v will be supporting to $|\cdot|$ at u, v everywhere in $[0, t_1]$. Let us make another transformation of the multipliers m_1, m_2 and the support functions n_u, n_v. Let us show that by changing these quantities one can obtain that everywhere on $[0, t_1]$ the following conditions are satisfied:

$$\begin{aligned} &\psi_1 = mr + n_u + g_1'^* m_1 , \qquad \psi_2 = n_v - mr + g_2'^* m_2 , \\ &g_1' n_u = 0 , \qquad g_2' n_v = 0 , \\ &|u| + |v| = 1 , \qquad (r, u-v) = 0 , \\ &(\psi_1, u) + (\psi_2, v) = 1 . \end{aligned} \tag{8}$$

Since $u = |u|n_u$, $v = |v|n_v$, then by (7) the relations (8) can be violated only at points where $u(t) = 0$ or $v(t) = 0$. For definiteness, let $u(t) = 0$ at t. Then $v(t) \neq 0$ and hence $g_2' n_v = 0$. Therefore, it remains for us to choose $\tilde{n}_u$, $\tilde{m}_1$ so that $\psi_1 = \tilde{n}_u + mr + g_1'^* \tilde{m}_1$, $g_1' \tilde{n}_u = 0$ and $\tilde{n}_u$ is supporting to $|\cdot|$ at 0. Note that if $\tilde{n}_u$ is supporting at 0, then $|\tilde{n}_u| \leq 1$. It is not hard to see that taking $\tilde{n}_u = \psi_1 - mr - g_1'^* \tilde{m}_1$ where m_1 yields $\min_{m_1} |\psi_1 - mr - g_1'^* \tilde{m}_1|^2$, we get the desired $\tilde{n}_u$, $\tilde{m}_1$. Indeed,

$$|\tilde{n}_u|^2 = \min_{\tilde{m}_1} |\psi_1 - mr - g_1'^* \tilde{m}_1|^2$$

$$\leq |\psi_1 - mr - g_1'^* m_1|^2 \leq 1 .$$

On the other hand, since $\tilde{m}_1$ is stationary for $|\psi_1 - mr - g_1'^* \tilde{m}_1|^2$, then $g_1'(\psi_1 - mr - g_1'^* \tilde{m}_1) = 0$. Whence $g_1' \tilde{n}_u = 0$. The assertion is proved.

Note that since we are changing m_1, m_2 only at points where $u(t) = 0$, $v(t) = 0$, respectively, the conditions (6) will be satisfied for the new n_u, n_v, m, m_1, m_2.

We now set about analyzing relations (8) with the goal of determining u, v, n_u, n_v, m, m_1, m_2. From the first equations we get

$$n_u = \psi_1 - mr - g_1'^* m_1 , \qquad n_v = \psi_2 + mr - g_2'^* m_2 .$$

According to (8),

$$n_u \in V_1(x) , \qquad n_v \in V_2(y)$$

and since

$$g_1'^* m_1 \perp V_1(x) , \qquad g_2'^* m_2 \perp V_2(y) ,$$

then

$$n_u = \bar{\psi}_1 - m\bar{r}_1 \quad , \qquad n_v = \bar{\psi}_2 + m\bar{r}_2 \quad ,$$

where the bar denotes projection onto $V_1(x)$, $V_2(y)$, respectively. Substituting the expressions for n_u, n_v into $g_1' n_u = 0$, $g_2' n_v = 0$ gives:

$$\begin{aligned} m_1 &= (g_1' \circ g_1'^*)^{-1} \circ g_1'(\psi_1 - mr) \quad ; \\ m_2 &= (g_2' \circ g_2'^*)^{-1} \circ g_2'(\psi_2 + mr) \quad . \end{aligned} \tag{10}$$

Since $u = |u| n_u$, $v = |v| n_v$, then from $(r, u-v) = 0$ and (9) we deduce

$$m = \frac{|u|(\bar{\psi}_1, \bar{r}_1) - |v|(\bar{\psi}_2, \bar{r}_2)}{|u|\,|\bar{r}_1|^2 + |v|\,|\bar{r}_2|^2} \quad . \tag{11}$$

Now let us establish some important properties of $u(t)$, $v(t)$. By (8) and (9), the following relations hold for $u = u(t)$, $v = v(t)$ for all $t \in [0, t_1]$:

$$\begin{aligned} &(n_u(t), u) + (n_v(t), v) = 1 \quad , \\ &|u| + |v| = 1 \quad , \\ &u = |u| n_u(t) \quad , \qquad v = |v| n_v(t) \quad , \\ &(r(t), u-v) = 0 \quad . \end{aligned} \tag{12}$$

Let us see for which t the vectors u, v are uniquely defined by this system of equations. Thus, consider a point t and the pair u, v satisfying (12).

First consider the case $|n_u(t)| < 1$. From (12) we conclude that $|u| = 0$, $|v| = 1$, whence $u = 0$, $v = n_v(t)$.

Thus u and v are uniquely determined from (12) for such a t. The case $|n_v(t)| < 1$ is analogous. Now consider t where one of (n_u,r), (n_v,r) is nonzero. Substituting $u = |u|n_u$, $v = |v|n_v$ into the equation $(r,u-v) = 0$ we get

$$|u|((n_u,r) + (n_v,r)) = (n_v,r) ,$$

$$|v|((n_u,r) + (n_v,r)) = (n_u,r) .$$

Since $|(n_u,r)| + |(n_v,r)| > 0$ it follows from these equations that $(n_u,r) + (n_v,r) \neq 0$. Hence,

$$|u| = (n_v,r)((n_u,r) + (n_v,r))^{-1} ,$$

$$|v| = (n_u,r)((n_u,r) + (n_v,r))^{-1} ,$$

whence u, v are uniquely defined and expressed by the formulas:

$$u = \frac{(n_v,r)}{(n_u,r) + (n_v,r)} n_u , \qquad v = \frac{(n_u,r)}{(n_u,r) + (n_v,r)} n_v .$$

We call points where $|n_u| = |n_v| = 1$:

$$(n_u,r) = (n_v,r) = 0 , \tag{13}$$

"degenerate" points.

From the above arguments it follows that at nondegenerate points u, v are defined by (12) uniquely. Using (9) it is easy to represent degeneracy conditions in the form:

$$\begin{gathered} |\bar{r}_1|^2(\bar{\psi}_2,\bar{r}_2) + |\bar{r}_2|^2(\bar{\psi}_1,\bar{r}_1) = 0 , \\ (|\bar{r}_1|^2 - |\bar{r}_2|^2) \frac{(\bar{\psi}_1,\bar{r}_1)^2}{|\bar{r}_1|^4} = |\bar{\psi}_1|^2 - |\bar{\psi}_2|^2 . \end{gathered} \tag{14}$$

From (14) it follows that the set of nondegenerate points is open. Further, since u, v are defined on it uniquely by (12), we see that $u(t)$, $v(t)$ are continuous on this set.

Lemma 1. $m \equiv$ const. on the intervals $\delta \subset \{t \mid t$ is nondegenerate, $|u||v| > 0\}$.

Proof. Let us first show that m is locally Lipschitzian on δ, for which we calculate m. Note that for $t \in \delta$ one has $|n_u| = |n_v| = 1$. Whence, since t is nondegenerate, $(n_u, r) \neq 0$ or $(n_v, r) \neq 0$. Now from the relations

$$|u|((n_u, r) + (n_v, r)) = (n_v, r) \quad ;$$

$$|v|((n_u, r) + (n_v, r)) = (n_u, r)$$

we derive the only possible cases:

- a. $(n_u, r) > 0$, $(n_v, r) > 0$,
- b. $(n_u, r) < 0$, $(n_v, r) < 0$, implying by (9) that

$$-\frac{(\bar\psi_2, \bar r_2)}{|\bar r_2|^2} < m < \frac{(\bar\psi_1, \bar r_1)}{|\bar r_1|^2}$$

and

$$\frac{(\bar\psi_1, \bar r_1)}{|\bar r_1|^2} < m < -\frac{(\bar\psi_2, \bar r_2)}{|\bar r_2|^2} ,$$

respectively.

Note that in case (a.), $A > 0$ while in case (b.), $A < 0$, where

$$A = |\bar r_1|^2(\bar\psi_2, \bar r_2) + |\bar r_2|^2(\bar\psi_1, \bar r_1) .$$

Let us calculate m. From (9) it follows that

$$|\bar{\psi}_1 - m\bar{r}_1|^2 = |\bar{\psi}_2 + m\bar{r}_2|^2 .$$

Whence

$$m^2(|\bar{r}_1|^2 - |\bar{r}_2|^2) - 2mb + (|\bar{\psi}_1|^2 - |\bar{\psi}_2|^2) = 0 ,$$

where $b = (\bar{\psi}_1, \bar{r}_1) + (\bar{\psi}_2, \bar{r}_2)$. Therefore, if $|\bar{r}_1| = |\bar{r}_2|$ at t, then

$$m = \frac{|\bar{\psi}_1|^2 + |\bar{\psi}_2|^2}{2b} . \tag{16}$$

But if $|\bar{r}_1| \neq |\bar{r}_2|$ then

$$m = \frac{b + \varepsilon\sqrt{\Delta}}{|\bar{r}_1|^2 + |\bar{r}_2|^2} ,$$

where

$$\varepsilon = \pm 1 , \quad \Delta = b^2 - (|\bar{r}_1|^2 - |\bar{r}_2|^2)(|\bar{\psi}_1|^2 - |\bar{\psi}_2|^2) .$$

Using (15) we find that $\varepsilon = -$ sign A and $\Delta > 0$ on δ. Thus for $|r_1| \neq |r_2|$,

$$m = \frac{b - \text{sign } A\sqrt{\Delta}}{|\bar{r}_1|^2 - |\bar{r}_2|^2} . \tag{17}$$

Now let $|\bar{r}_1| = |\bar{r}_2|$ at $t \in \delta$. Then $A|_t = |\bar{r}_1|^2 b_t$. Take a neighborhood U_t where $Ab > 0$. Then in U_t formulas (16), (17) can be combined:

$$m = \frac{|\bar{\psi}_1|^2 - |\bar{\psi}_2|^2}{b + \text{sign } A\sqrt{\Delta}} ,$$

i.e., m is Lipschitzian in U_t. But if $|\bar{r}_1| \neq |\bar{r}_2|$ at t, then we take a neighborhood U_t, where $|\bar{r}_1| \neq |\bar{r}_2|$. Then in U_t m has the form (17) and so is Lipschitzian. Thus m is locally Lipschitzian on δ.

Now let us show that $\dot{m} \equiv 0$ on δ. From (9), (10) we conclude that n_u, n_v, m_1, m_2 are locally Lipschitzian on δ. Now substituting 6.2 into 6.3 we get $\dot{n}_u = -\dot{m}r - g_1'{*}\dot{m}_1$. Whence

$$\dot{m}(r,u) = -(\dot{n}_u,u) = -\frac{|u|}{2}\frac{d}{dt}|n_u|^2 .$$

The right side of this relation is identically zero on δ. Only for points with $|u| > 0$ this is not obvious. But where $|n_u|^2 = 1 = \max_{\delta} |n_u|^2$, whence $\frac{d}{dt}|n_u|^2 = 0$. Thus $\dot{m}(r,u) \equiv 0$ on δ. Since $(r,u) = |u|(r,n_u)$ and $|u| > 0$, $(r,n_u) \neq 0$, then $(r,u) \neq 0$ and $\dot{m} \equiv 0$. Whence $m \equiv$ const. on δ.

Theorem 1. The functions n_u, n_v, m, m_1, m_2 are Lipschitzian.

Proof. By (9), (10) it suffices to show m is Lipschitzian. For this it suffices to prove that the derivatives of the Dini number m, say, the right ones, are uniformly bounded.

Let t be degenerate. Then from (14)

$$\frac{(\bar{\psi}_1,\bar{r}_1)}{|\bar{r}_1|^2} = -\frac{(\bar{\psi}_2,\bar{r}_2)}{|\bar{r}_2|^2}$$

and so for any t'

$$m(t) = \frac{|u|_{|t'}(\bar{\psi}_1,\bar{r}_1)_{|t} - |v|_{|t'}(\bar{\psi}_2,\bar{r}_2)_{|t}}{|u|_{|t'}|\bar{r}_1|^2_t + |v|_{|t'}|\bar{r}_2|^2_t} .$$

From this formula we easily deduce that $\frac{m(t')-m(t)}{t'-t}$ and the derivatives of m are uniformly bounded at degenerate points.

Let t be nondegenerate. The following cases are possible.

- 1. There exists $t'' > t$ such that on (t,t'') the points are nondegenerate and $|u||v| > 0$. Then by Lemma 1, $m \equiv$ const. on (t,t''). At the same time u, v are continuous in t. Hence by (11) m is continuous in t and $m \equiv \text{const.}$ on $[t,t'')$, whence $\dot{m}_t = 0$.

● 2. For any $t'' > t$ such that on $[t,t'']$ the points are nondegenerate there are points on $(t,t'']$, where $|u||v| = 0$. Then by the continuity of u, v in t one has $|u||v|_{|t} = 0$. For definiteness, let $u(t) = 0$. But then, again from the continuity of u, v on $[t,t'']$ we deduce that for small $t''-t$ one has on $[t,t'']$ that $u = 0$ for $|u||v| = 0$.

Let us estimate

$$\frac{m(t') - m(t)}{t' - t} \qquad \text{for} \quad t < t' \leq t'' \quad .$$

If $u(t') = 0$ then

$$m(t') = - \left. \frac{(\bar{\psi}_2, \bar{r}_2)}{|\bar{r}_2|^2} \right|_{t'}$$

and since

$$m(t) = - \left. \frac{(\bar{\psi}_2, \bar{r}_2)}{|\bar{r}_2|^2} \right|_{t} ,$$

then

$$\frac{m(t') - m(t)}{t' - t} = \frac{1}{t' - t} \Delta_{t'-t} \left\{ - \frac{(\bar{\psi}_2, \bar{r}_2)}{|\bar{r}_2|^2} \right\} . \tag{18}$$

If $|u||v|_{|t'} > 0$ then there exists

$$t^* = \max \{\tilde{t} \mid \tilde{t}<t',\ u(\tilde{t})=0\} \quad .$$

On $[t^*,t']$ one has $|u||v| > 0$ and hence from Lemma 1 $m \equiv \text{const.}$ on $[t^*,t']$. Thus

$$\left|\frac{m(t') - m(t)}{t' - t}\right| = \left|\frac{m(t^*) - m(t)}{t' - t}\right| \leq \left|\frac{m(t^*) - m(t)}{t^* - t}\right|$$
$$= \frac{1}{t^* - t}\,\Delta_{t^*-t}\left\{-\frac{(\bar{\psi}_2, \bar{r}_2)}{|\bar{r}_2|^2}\right\}. \tag{19}$$

Since (18), (19) are uniform estimates, the proof is finished.

§ 4. STATIONARY SETS

Now let us give the stationary conditions a more concise form. Let us prove the Assertion: In order that the trajectory x, y be an extremal, it is necessary and sufficient that there exist a set of functions n_u, n_v, p, p_1, p_2, among which n_u, n_v, p_1, p_2 are vector-valued in E^n, E^n, E^{k_1}, E^{k_2}, respectively, and p is a scalar function, n_u, n_v are Lipschitzian, and p, p_1, p_2 are bounded measurable functions such that the following conditions hold:

- a. n_u, n_v are almost everywhere supporting to $|\cdot|$ at the points $\dot{x}$, $\dot{y}$;
- b. $|x-y| = 1$, $|\dot{x}| + |\dot{y}| = 1$, $g_1(x) = x_1'(x)n_u = 0$,

$$g_2(y) = g_2'(y)n_v = 0 \quad ; \tag{20}$$

- c. $\dot{n}_u = -pr - g_1'^*(x)p_1$, $\dot{n}_v = pr - g_2'^*(y)p_2$.

Proof. Necessity. If x, y is an extremal, there exists a set of conjugate multipliers, ψ_1, ψ_2, m, m_1, m_2 and support functions n_u, n_v to $|\cdot|$ at the points $\dot{x}$, $\dot{y}$ such that conditions (16) are satisfied and n_u, n_v, m, m_1, m_2 are Lipschitzian (Theorem 1). Setting $p = \dot{m}$, $p_1 = \dot{m}_1$, $p_2 = \dot{m}_2$, we get the desired n_u, n_v, p, p_1, p_2.

Sufficiency. Converse. Suppose such a set exists. Set

$$m,\ m_1,\ m_2\Big|_t = m',\ m_1',\ m_2' - \int_t^{t_1} p, p_1, p_2\ d\tau \quad ,$$

where $m',\ m_1',\ m_2'$ are as yet undefined quantities. Form the functions

$$\psi_1 = n_u + mr + g_1'^* m_1 \quad , \qquad \psi_2 = n_v - mr + g_2'^* m_2 \quad .$$

This set of multipliers obviously satisfies all conditions (6) except maybe the boundary condition. To satisfy that, i.e.,

$$(\psi_1(t_1),\ \psi_2(t_1)) \in L$$

or

$$(n_u^1, n_v^1) + G'^*(m', m_1', m_2') \in \text{Ker } G' \quad ,$$

we need to set

$$m',\ m_1',\ m_2' = -(G' \circ G'^*)^{-1} \circ G'(n_u', n_v') \quad .$$

Now let us introduce a notion that is essential for us. We will call the set of functions $x,\ y,\ n_u,\ n_v,\ p,\ p_1,\ p_2$ a stationary set or regime if conditions (20) hold for it.

As we have established, extremals are none other than the $x,\ y$ projections of stationary sets. Consequently, we can get a description of all extremals by describing all stationary sets.

We will call a set of vectors $x_0,\ y_0,\ n_u^0,\ n_v^0$ degenerate if

$$g_1(x_0) = g_1'(x_0)n_u^0 = 0 \ , \qquad g_2(y_0) = g_2'(y_0)n_v^0 = 0 \ ,$$

$$|r_0| = |n_u^0| = |n_v^0| = 1 \ ,$$

$$(n_u^0, r_0) = (n_v^0, r_0) = 0 \ .$$

The above differential equations for n_u, n_v can be briefly written

$$\frac{d}{dt}(n_u, n_v) = -G'^*(p, p_1, p_2) .$$

This highlights the apparent geometric significance of extremality in Ulam's problem: the trajectory x, y is an extremal if and only if when moving along it the derivative of the support function $n_{uv}(t) = (n_u(t), n_v(t))$ assumes values orthogonal to the tangent plane of the surface $G(x,y) = 0$. We may interpret Ulam's problem as the problem of drawing the shortest curve $x(t)$, $y(t)$ on the surface $G(x,y) = 0$ joining the points (x_0^*, y_0^*), (x_1^*, y_1^*), where length of a curve is measured in the nonsmooth norm $\|(x,y)\| = |x| + |y|$. In this interpretation the differential equation corresponds to an equation for geodesics in classical differential geometry. However, whereas in classical differential geometry the smoothness of the support function is self-evident, in Ulam's problem to prove that n_u, n_v are Lipschitzian we have had to make a special study. This study was made under the assumption of general positions of the segment on the extremal. But if on the trajectory in question x, y not all positions of the segment satisfy the general position condition $|\bar{r}_1||\bar{r}_2| > 0$, then, generally speaking, the support function $n_{uv}(t)$ becomes discontinuous and equations (20) cease to be an equivalent way of expressing extremality.

EXAMPLE. Consider Ulam's problem on E^2 for the case when the surfaces M_1, M_2 are a pair of nonparallel lines. Let the boundary positions of the segment x_0^*, y_0^*; x_1^*, y_1^* be such that

$$(r_0^*,\ x_0^*-x_1^*)(r_1^*,\ x_0^*-x_1^*) < 0 , \qquad (r_0^*,\ y_0^*-y_1^*)(r_1^*,\ y_0^*-y_1^*) < 0 .$$

Then in moving the segment from initial position to final position we can find two instants of time when the segment becomes orthogonal to the line M_1 or M_2. At those instants the support function $n_{uv}(t)$ is discontinuous.

Let us agree on some terminology convenient for the sequel. We will call the set of functions $x, y, n_u, n_v, p, p_1, p_2$ a stationary set or a regime if conditions (20) are satisfied.

As we have just shown, extremals in Ulam's problem are no more than the x, y projections of stationary sets. And, it is convenient to classify extremals in terms of stationary sets. By describing all possible sets we will have a description of all extremals.

We will call the set of vectors x_0, y_0, n_u^0, n_v^0 degenerate if

$$g_1(x_0) = g_1'(x_0)n_u^0 = 0 , \qquad g_2(y_0) = g_2'(y_0)n_v^0 = 0 ,$$

$$|r_0| = |n_u^0| = |n_v^0| = 1 ,$$

$$(n_u^0, r_0) = (n_v^0, r_0) = 0 .$$

§ 5. OPERATION OF PARALLEL DISPLACEMENT OF VECTORS ON SURFACES AND EQUATIONS (20) OF A STATIONARY SET

Equations (20) can be written in a more concise form. For each τ and $t \in \Delta$ we denote by $N_u(\tau,t)$ the vector obtained from $n_u(t)$ by a parallel displacement along the curve x from the point $x(t)$ to the point $x(\tau)$. We recall the formal definition of the operation of parallel displacement, well known in the Differential Geometry. By the parallel displacement of the vector on the surface M_1 along the curve x we mean a vector Lipschitzian $\ell(\cdot)$ such that

$$\ell(t) \in V_1(x(t)) \ , \qquad \dot{\ell}(t) \in V_1^{\perp}(x(t)) \ .$$

By definition, ℓ satisfies the linear differential equation

$$\dot{\ell} = -g_1'^{*} \circ (g_1' \circ g_1'^{*})^{-1} g_1'' \dot{x}, \ \ell \ .$$

The function $N_u(\tau,t)$ can be written as $N_u(\tau,t) = X_1(t,\tau)n_u(t)$, where X_1 is the fundamental matrix of the differential equation for the parallel displacement. Since X_1 is Lipschitzian, N_u is also Lipschitzian in the variables τ and t. Next, let E denote the subset of full measure in Δ, consisting of points of the approximate continuity of the measurable functions $\dot{x}$ and $\dot{n}_u$. It is easy to show that Lipschitzians in X_1 and N_u are differentiable at the points of the set $E \times E$. Since $N_u(t,t) = n_u(t)$, we have for any $t \in E$

$$\dot{n}_u(t) = \frac{\partial}{\partial t} N_u(t,t) + \frac{\partial}{\partial \tau} N_u(t,t) \ ,$$

and since

$$\frac{\partial}{\partial t} N_u(t,t) \in V_1(x(t)) \ ,$$

$$\frac{\partial}{\partial \tau} N_u(t,t) \in V_1^{\perp}(x(t)) \ ,$$

we obtain

$$\frac{\partial}{\partial t} N_u(t,t) = P_{V_1(x(t))} \dot{n}_u(t) \ .$$

Thus, if x, y, n_u, n_v, P, R_1, P_2 is a stationary set, for almost all $t \in \Delta$

$$\frac{\partial}{\partial t} N_u(t,t) = -p\bar{r}_1 \ , \qquad \frac{\partial}{\partial t} N_v(t,t) = p\bar{r}_2 \ .$$

This system of equations is equivalent to the system (20).

It is possible to single out the cases where this system of d ferential equations possesses the linear first integral. Suppose the surfaces coincide in our problem: $M_1 = M_2 = M$. We choose any $a < t_0$, where t_0, t_1 are the endpoints of Δ and continue x, to the segment $[a,t_0]$ in any arbitrary way keeping $x(a) = y(a)$. We also consider the functions $\bar{n}_u(t) = N_u(0,t)$, $\bar{n}_v(t) = N_v(0,t)$ obtained by displacing $n_u(t)$, $n_v(t)$ along the Lipschitzian curve x, y to the point x(a). Then

$$\dot{\bar{n}}_u(t) = X_1(t,0)\frac{\partial}{\partial t} N_u(t,t) = -pX_1(t,0)\bar{r}_1 \quad ,$$

$$\dot{\bar{n}}_v(t) = X_2(t,0)\frac{\partial}{\partial t} N_v(t,t) = pX_2(t,0)\bar{r}_2 \quad ,$$

$$\dot{\bar{n}}_u(t) + \dot{\bar{n}}_v(t) = p(X_2(t,0)\bar{r}_2 - X_1(t,0)\bar{r}_1) \quad .$$

Thus, if the surface M is such that the result of the parallel displacement does not depend on the mode of displacement and, in a dition, for any $x, y \in M$ under the parallel displacement from x to y the $\bar{r}_1$ becomes $\bar{r}_2$, where $r = x - y$, then $\bar{n}_u(t) + \bar{n}_v(t$ const, i.e., the sum of n_u and n_v displaced to a single point of the surface is constant. This invariant plays a crucial role i solving the problems in which the constraint surfaces possess the stated property. Regrettably, the class of such surfaces is narro they include only subspaces and cylinders $S^1 \times R^k$.

§ 6. STANDARD REGIMES

Let us examine a stationary set defined on the segment Δ. Introduce the sets

$$G = \{t \in \Delta \mid t \text{ is nondegenerate}\} \;,$$

$$F = \{t \in G \mid (r,u) \neq 0\} \;.$$

Since G is open and u, v are continuous on G, then F is also open. At each point on $G \setminus F$, $u = 0$ or $v = 0$. Indeed, suppose not, i.e., there is $t \in G \setminus F$ where $|u||v| > 0$. Then $|n_u| = |n_v| = 1$ and since t is nondegenerate then $(n_u, r) \neq 0$. But then $(r,u) = |u|(r,n_u) \neq 0$, a contradiction. On the component intervals of F, $m \equiv$ const. (Lemma 1), and therefore $p = \dot{m} = 0$.

We see that it is natural to isolate the following types of regimes:

- a. a regime nondegenerate in $\operatorname{int}\Delta$ and with $(r,u) \neq 0$ in $\operatorname{int}\Delta$;
- b. a regime nondegenerate in $\operatorname{int}\Delta$ for which $u \equiv 0$ or $v \equiv 0$ on Δ;
- c. a regime for which all points $t \in \Delta$ are degenerate.

For a regime (a.), $\operatorname{int}\Delta \subset F$, therefore $p \equiv 0$. Equation (20(c.)) in this case takes the form

$$\dot{n}_u = -g_1'^*(x)p_1 \;, \qquad \dot{n}_v = -g_2'^*(y)p_2 \;.$$

Consequently, if we parametrize x, y by the length covered by them then $\ddot{x} = g_1'^*\tilde{p}_1$, $\ddot{y} = g_2'^*\tilde{p}_2$, where $\tilde{p}_1$, $\tilde{p}_2$ are bounded measurable functions. Noting that these equations are the equations of the geodesics on M_1, M_2, we conclude that in this regime the endpoints of the segment move (slide) along the geodesics. It is natural to call such a regime a sliding regime.

For the regime (b.), $x \equiv$ const. or $y \equiv$ const. and there occurs, so to speak, a rotation of the segment about a fixed endpoint. Hence we call such a regime a rotating regime.

For these regimes one can give a representation in the form of a system of differential equations. Since in solving specific problems it is appropriate to use representations of standardized regimes, we shall derive the corresponding formulas.

a. A Sliding Regime

Let the geodesics along which the endpoints of the segment move be $q_1(\alpha)$, $q_2(\beta)$, and for parameters α, β we take the length of the corresponding arcs. From equations (20) we conclude that the regime is defined by the numbers α_0, β_0, t_0 and the geodesics q_1, q_2 as follows:

$$x(t) = q_1(\alpha(t)) \quad , \qquad y(t) = q_2(\beta(t)) \quad ,$$

$$n_u(t) = \dot{q}_1(\alpha(t)) \quad , \qquad n_v(t) = \dot{q}_2(\beta(t)) \quad ,$$

$$p = 0 \quad , \qquad p_1 = -\frac{\gamma}{\sigma+\gamma}(g_1' \circ g_1'^*)^{-1} \circ g_1'\ddot{q}_1(\alpha) \quad ,$$

$$p_2 = -\frac{\sigma}{\sigma+\gamma}(g_2' \circ g_2'^*)^{-1} \circ g_2'\ddot{q}_2(\beta) \quad ,$$

where

$$\sigma = (\dot{q}_1(\alpha),\ q_1(\alpha) - q_2(\beta)) \quad ,$$

$$\gamma = (\dot{q}_2(\beta),\ q_1(\alpha) - q_2(\beta)) \quad ,$$

and α, β is a solution of the system

$$\dot{\alpha} = \frac{\gamma}{\sigma+\gamma} \quad , \qquad \dot{\beta} = \frac{\sigma}{\sigma+\gamma} \quad ,$$

$$\alpha(t_0) = \alpha_0 \quad , \qquad \beta(t_0) = \beta_0$$

and $\sigma\gamma > 0$ in $\operatorname{int} \Delta$.

b. A Rotating Regime

Representation of the rotating regime is obtained in the same way. The regime of rotation about "x" is defined by x_0, y_0, n_u^0, n_v^0, t_0, where

$$|n_u^0| \leq |n_v^0| = |r_0| = 1 , \qquad (n_v^0, r_0) = 0 ,$$

$$g_1'(x_0)n_v^0 = 0 , \qquad g_2'(y_0)n_v^0 = 0 .$$

It has the form:

$$x \equiv x_0, y, n_u, n_v, p = \frac{1 + (r,\ g_2'^* \circ (g_2' \circ g_2'^*)^{-1} \circ g_2'' n_u,\ n_v)}{|\bar{r}_2|^2} ,$$

$$p_1 = -p(g_1' \circ g_1'^*)^{-1} \circ g_1' r ,$$

$$p_2 = (g_2' \circ g_2'^*)^{-1}(g_2'' n_v,\ n_v + p g_2' r) ,$$

where y, n_u, n_v is a solution of the system

$$\dot{y} = n_v , \qquad \dot{n}_u = -p\bar{r}_1 , \qquad \dot{n}_v = pr - g_2'^*(y)p_2$$

subject to the boundary condition

$$y(t_0) = y_0 , \qquad n_u(t_0) = n_u^0 ,$$

$$n_v(t_0) = n_v^0 .$$

Here one has $|n_u| \leq 1$ in $\operatorname{int} \Delta$ and there is no degeneracy.

c. A Degenerate Regime

Here we obtain that in order that the set x, y, n_u, n_v, p, p_1, p_2 be degenerate, it is necessary and sufficient that there exist measurable weights α, β such that $p = f_1 = f_2$, $p_1 = \tilde{p}_1$, $p_2 = \tilde{p}_2$,

$$f_1 = \frac{\alpha(1 - (r,\ g_1'^{*}\circ(g_1'\circ g_1'^{*})^{-1}\circ g_1''n_u,\ n_u)) - \beta(n_u, n_v)}{|\bar{r}_1|^2},$$

where

$$f_2 = \frac{\beta(1 + (r,\ g_2'^{*}\circ(g_2'\circ g_2'^{*})^{-1}\circ g_2''n_v,\ n_v)) - \alpha(n_u, n_v)}{|\bar{r}_2|^2},$$

$$\tilde{p}_1 = (g_1\circ g_1'^{*})^{-1}(\alpha g_1''n_u,\ n_u - f_1 g_1' r) ,$$

$$\tilde{p}_2 = (g_2'\circ g_2'^{*})^{-1}(\beta g_2''n_v,\ n_v - f_2 g_2' r)$$

and x, y, n_u, n_v is a solution of the system

$$\dot{x} = \alpha n_u , \qquad \dot{y} = \beta n_v ,$$

$$\dot{n}_u = -f_1 r - g_1'^{*}\tilde{p}_1 , \qquad \dot{n}_v = f_2 r - g_2'^{*}\tilde{p}_2 ,$$

and the set $x(t_0)$, $y(t_0)$, $n_u(t_0)$, $n_v(t_0)$ is degenerate for some $t_0 \in \Delta$.

We set out the condition

$$f_1 = f_2 \tag{21}$$

for special references.

The solution of Ulam's problem has thus been reduced to studying the selected standard regimes and their mutual conjuction.

From the representations of sliding and rotation we see that these regimes can be uniquely extended to a maximal interval. We call a regime on a maximal interval a complete regime. If on Δ a regime contains a section of sliding and δ is the interval of the corresponding complete sliding, then on $\delta \cap \Delta$ the original regime is a sliding. In fact, let $\tilde{x}$, $\tilde{y}$, $\tilde{n}_u$, $\tilde{n}_v$, $\tilde{p}$, $\tilde{p}_1$, $\tilde{p}_2$ be sliding on δ extending the existing sliding section. If

$$\tilde{x},\tilde{y},\tilde{n}_u,\tilde{n}_v,\tilde{p},\tilde{p}_1,\tilde{p}_2 \neq x,y,n_u,n_v,p,p_1,p_2 \mid \delta \cap \Delta ,$$

then there exists $t^* \in \delta \cap \Delta$, a boundary point of the set of coincidence. Then

$$\tilde{x},\tilde{y},\tilde{n}_u,\tilde{n}_v\big|_{t^*} = x,y,n_u,n_v\big|_{t^*}$$

by continuity and therefore in a neighborhood of t^*, one has that x, y, n_u, n_v, p, p_1, p_2 is nondegenerate and $(r,u) \neq 0$. Hence by definition, in this neighborhood our regime is a sliding and coincides with $\tilde{x}$, $\tilde{y}$, $\tilde{n}_u$, $\tilde{n}_v$, $\tilde{p}$, $\tilde{p}_1$, $\tilde{p}_2$, a contradiction.

In the particular problems which we shall examine, every stationary set consists of a finite number of component standard regimes. In the general case, of course, such sections of a regime can be stacked up in arbitrarily complex fashion.

• • • • •

CHAPTER 2

OPTIMAL MOTION OF SEGMENTS IN THE n-DIMENSIONAL SPACE AND ON THE n-DIMENSIONAL SPHERE

We combine both problems by viewing space as a sphere of infinite radius. For this, as endpoint constraints we take

$$g_1(x) = g_2(x) = g(x) = \rho|x|^2 + (\ell, x) ,$$

where

$$\rho \geq 0 , \qquad |\ell| = 1 .$$

Then for $\rho = 0$ these constraints determine a subspace of dimension $n-1$, and for $\rho > 0$ a sphere of radius $\frac{1}{2\rho}$ centered at $-\frac{\ell}{2\rho}$. Obviously, a unit segment can be placed with endpoints on the surface $M = \{x \mid g(x) = 0\}$ only for $\rho \leq 1$. On the other hand, $|\bar{r}_1| = |\bar{r}_2| = \sqrt{1-\rho^2}$. Hence general position of a segment with endpoints on M will hold only for $\rho < 1$. We will therefore consider the case $0 \leq \rho < 1$ (the case $\rho = 1$ beyond the scope of our analysis is not interesting: it is the problem of optimal motion of unit segments on a sphere of unit diameter).

Below we will often use the notation: $s = 1 - 2\rho^2$, $c = \sqrt{1-\rho^2}$. Let us study standard regimes for our problem.

§ 1. THE SLIDING REGIME

We study a complete sliding. From the general representation of sliding we see that it is defined by variation of the variables $\sigma = (n_u, r)$, $\gamma = (n_v, r)$. In our problem it is easily verified that σ, γ, $\theta = (n_u, n_v)$ satisfy the closed system of differential equations:

$$\dot{\sigma} = \frac{\gamma s - \sigma\theta}{\sigma + \gamma}, \qquad \dot{\gamma} = \frac{\gamma\theta - \sigma s}{\sigma + \gamma}, \qquad \dot{\theta} = 4\rho^2(\sigma - \gamma),$$

and in the sliding interval we must have $\sigma\gamma > 0$. It is convenient to change from the variables σ, γ, θ to the variables $\kappa = \sigma + \gamma$, $\chi = \sigma - \gamma$, θ. The system then takes the form:

$$\dot{\kappa} = -(\theta + s)\frac{\chi}{\kappa}, \qquad \dot{\chi} = s - \theta, \qquad \dot{\theta} = 4\rho^2\chi .$$

The condition $\sigma\gamma > 0$ then becomes $|\kappa| > |\chi|$, and $\kappa = \chi = 0$ becomes the degeneracy criterion.

Let us integrate the derived system:

if $\rho = 0$ then

$$\kappa^2 = \kappa_0^2 + (\theta^2 - 1)(t - t_0)^2 - 2(1+\theta)\chi_0(t - t_0),$$

$$\chi = \chi_0 + (1-\theta)(t - t_0), \qquad \theta \equiv \theta_0 ;$$

if $\rho > 0$ then

$$\chi = \frac{A}{2\rho}\cos(2\rho t + \phi),$$

$$\kappa^2 = \kappa_0^2 - \frac{2s + A\sin(2\rho t+\phi) - (\theta_0+s)^2}{2\rho^2} , \qquad (22)$$

$$\theta = s - A\sin(2\rho t+\phi) ,$$

where

$$A = \sqrt{4\rho^2\chi_0^2 + (\theta_0-s)^2} ,$$

$$\phi = \arg(2\rho\chi_0 + (\theta_0-s)i - 2\rho t_0) .$$

The interval of complete sliding extending the given one is the greatest interval containing the original one, inside which $\kappa^2 - \chi^2 > 0$. Simple analysis of formulas (22) leads us to the conclusions:

- 1. if

$$4\rho^2(\kappa_0^2 - \chi_0^2) + s(\theta_0 - s) - A|s| > 0 ,$$

then complete sliding is defined on R;

- 2. if

$$4\rho^2(\kappa_0^2 - \chi_0^2) + s(\theta_0 - s) - A|s| = 0 ,$$

then $s \neq 0$ and complete sliding is defined on a segment of length $\frac{\pi}{\rho}$ for $\rho > 0$ and on R for $\rho = 0$, and degenerates on the boundary.

- 3. if

$$4\rho^2(\kappa_0^2 - \chi_0^2) + s(\theta_0 - s) - A|s| < 0 , \qquad (23)$$

then $s \neq 0$, complete sliding is defined on a segment with length depending continuously on $|\kappa_0|$, $|\chi_0|$, θ_0, and does not reach degeneracy on the boundary.

We call such a sliding a simple sliding.

• 4. In cases (2.), (3.), if $[\alpha, \beta]$ is a segment of complete sliding then

$$x_\alpha = x_\beta \qquad \chi_\alpha = -\chi_\beta \qquad \theta_\alpha = \theta_\beta$$

Note that in the spatial case, i.e., for $\rho = 0$,

$$4\rho^2(\kappa_0^2 - x_0^2) + s(\theta_0 - s) - A|s| = (\theta_0 - 1) - |\theta_0 - 1|$$

$$\leq 0 ,$$

and hence only cases (2.) and (3.) are realized. In case (3.), $-1 < \theta_0 = (n_u^0, n_v^0) < 1$, i.e., the directional vectors of the geodesics (lines) are not parallel and the sliding goes along crossing lines. In case (2.), $\theta_0 = (n_u^0, n_v^0) = 1$, whence $n_u^0 = n_v^0 = d$. Since $\dot{n}_u \equiv 0$, $\dot{n}_v \equiv 0$, the sliding representation gives us:

$$\dot{\alpha} = \frac{1}{2} , \qquad \dot{\beta} = \frac{1}{2} ,$$

whence

$$x = x_0 + \int_{t_0}^{t} \frac{d}{2}\, d\tau = x_0 + \frac{t-t_0}{2}\, d ,$$

$$y = y_0 + \frac{t-t_0}{2}\, d ,$$

i.e., the regime is a parallel displacement (shift) of the segment.

For $\rho > 0$, i.e., on the sphere, as geodesics one has circles of large radius, and the segment endpoints slide along a pair of these circles.

It is also useful to describe the values of x, y, n_u, n_v on the boundary of complete regimes:

• 1. the set x_0, y_0, n_u^0, n_v^0 is the boundary of a simple complete sliding if and only if

$$\chi_0 s \neq 0 ,$$

$$g(x_0) = g(y_0) = g' n_u^0 = g' n_v^0 = 0 ,$$

$$|x_0| = |\chi_0| , \tag{24}$$

$$|r_0| = |n_u^0| = |n_v^0| = 1 .$$

Here $\chi_0 s < 0$ on the left endpoint and $\chi_0 s > 0$ on the right endpoint;

• 2. the set x_0, y_0, n_u^0, n_v^0 is the boundary of a complete sliding reaching degeneracy if and only if $s(\theta_0 - s) > 0$, and the set is degenerate. This boundary condition is two-sided.

§ 2. THE ROTATING REGIME

For definiteness, we consider rotation about the endpoint x. We will call the rotation complete if it is extended to a maximal interval.

From the general representation of a rotation it follows that the interval of complete rotation Δ is determined by the condition: $|n_u| < 1$ or $\sigma \neq 0$ in $\operatorname{int} \Delta$. It is easily checked that

$$\frac{d}{dt} |n_u|^2 = - \frac{2s}{c^2} \sigma ,$$

$$\dot{\sigma} = -\theta - s , \qquad \dot{\theta} = \frac{\sigma}{c^2} .$$

Let us integrate this system:

$$\sigma = A\cos\left(\frac{t}{c}+\phi\right), \qquad \theta = \frac{A}{c}\sin\left(\frac{t}{c}+\phi\right),$$
$$|n_u|^2 = |n_u^0|^2 + 2s(\theta_0+s) - 2\frac{s}{c}A\sin\left(\frac{t}{c}+\phi\right), \tag{25}$$

where

$$A = \sqrt{\sigma_0^2 + c^2(\theta_0+s)^2},$$

$$\phi = \arg\left(\sigma_0 + ic(\theta_0 - s)\right) - \frac{t_0}{c}.$$

Since

$$\max |n_u|^2 = |n_u^0|^2 + 2s(\theta_0+s) + \frac{2}{c}|s|A,$$

we get:

- 1. if

$$|n_u^0|^2 + 2s(\theta_0+s) + \frac{2A|s|}{c} < 1$$

then

$$\Delta = R;$$

- 2. if (26)

$$|n_u^0|^2 + 2s(\theta_0+s) + \frac{2A|s|}{c} = 1$$

then

$$|\Delta| = \begin{cases} \pi c & \text{if } s = 0 \\ 2\pi c & \text{if } s \neq 0 \end{cases}$$

and degeneracy is reached on the boundary;

- 3. if

$$|n_u^0|^2 + 2s(\theta_0+s) + \frac{2A|s|}{c} > 1,$$

then

$$|\Delta| < \pi c$$

and depends continuously on θ_0, $|n_u^0|$, σ_0. Degeneracy on the boundary is not attained. We say that such a rotation is simple;

• 4. Let $[\alpha,\beta]$ be a segment of complete rotation. Then in case (2.) for $s = 0$ one has $\theta_\alpha = -\theta_\beta$, and for $s \neq 0$ one has $\theta_\alpha = \theta_\beta$. But in case (3.),

$$\kappa_\alpha = -\kappa_\beta \qquad \chi_\alpha = -\chi_\beta \qquad \theta_\alpha = \theta_\beta .$$

Note that in the spatial case $(\rho = 0)$ rotation can be only simple, and on the sphere all cases are realized. Moreover, for $\rho = 0$ rotation of the segment occurs in a two-dimensional plane. Indeed, the representation of rotation gives us: $\dot{r} = -n_v$, $\dot{n}_v = \rho r$. This system is linear and has scalar coefficients. Hence r, n_v vary in some two-dimensional subspace and the segment $[x,y]$ rotates on a two-dimensional plane.

We complete the study of rotation by describing the values of x, y, n_u, n_v on the boundary of a complete regime.

From (25) we get:

• 1. x_0, y_0, n_u^0, n_v^0 is the boundary condition of a complete simple rotation if and only if

$$|x_0| = |\chi_0| ,$$

$$g(x_0) = g(y_0) = g'n_u^0 = g'n_v^0 = 0 ,$$

$$|n_u^0| = |n_v^0| = |r_0| = 1 , \tag{27}$$

$$\chi_0 s \neq 0 ,$$

and here $\chi_0 s > 0$ on the left endpoint and $\chi_0 s < 0$ on the right endpoint, for rotating about x, $\kappa_0\chi_0 > 0$, and for rotating about y, $\kappa_0\chi_0 < 0$;

• 2. x_0, y_0, n_u^0, n_v^0 is the boundary condition (right, left) of rotation (about x, y) attaining degeneracy if and only if this set is degenerate and $s = 0$, $\theta_0 \neq 0$ or $s(\theta_0 + s) < 0$.

§ 3. THE DEGENERATE REGIME

In this case condition (21) assumes the form

$$(\alpha-\beta)(\theta+s) = 0 \quad .$$

It is not hard to check also that $\dot{\theta} \equiv 0$, whence $\theta \equiv$ const. Hence it is natural to single out two types of degenerate regimes: a regime with $\theta = -s$ and any measurable weights α, β and also a regime with $\alpha = \beta = \frac{1}{2}$, $\theta \neq -s$. We call them a "weighted" regime and an ordinary regime, respectively.

For $\rho = 0$ a weighted degenerate regime is none other than a planar motion of the segment in which it rotates in one direction and the instantaneous center of rotation lies on it. The system of equations of an ordinary degenerate regime is easily integrated, and it can be described explicitly. We describe it for the spatial case since here it is geometrically transparent. From the representation of the degenerate regime we get:

$$p_1 = p_y \equiv 0 \quad ;$$

$$\dot{n}_u = -pr \;, \qquad \dot{n}_v = pr \;, \tag{28}$$

$$\dot{r} = \alpha n_u - \beta n_v = \frac{n_u - n_v}{2} \quad .$$

Set $d = n_u + n_v$ ($d \equiv$ const. since $\dot{d} \equiv 0$). Then $0 < |d| \leq 2$

since

$$|n_u| = |n_v| \equiv 1 ,$$

$$\theta = (n_u, n_v) \neq -s = -1 .$$

Let $\bar{n}_u$, n_u^d denote the components of the decomposition of n_u with respect to the orthogonal complement of d and the vector d. Then $\bar{n}_u = -\bar{n}_v$, $n_u^d + n_u^d = d$. By substituting these expressions into (28) we get

$$\dot{n}_u^d = \dot{n}_v^d = 0 , \qquad \dot{\bar{n}}_u = -pr , \qquad \dot{\bar{n}}_v = pr .$$

From the last equations it is seen that $\bar{n}_u$, r can be embedded in a two-dimensional subspace V orthogonal to d and that n_u^d, $n_u^d \equiv$ const. Noting that

$$|n_u^d| = \sqrt{|n_u|^2 - |\bar{n}_u|^2} = \sqrt{1 - |\bar{n}_u|^2} = \sqrt{1 - |\bar{n}_v|^2}$$
$$= |n_v^d| ,$$

we deduce from $n_u^d + n_v^d = d$ that $n_u^d = n_v^d \equiv \frac{d}{2}$. Take an orthonormal basis i, j in V. Then $r = e^{\phi j}$, where ϕ is Lipschitzian. Moreover, since

$$\dot{r} = \dot{\phi} j e^{\phi j} = \bar{n}_u ,$$

then

$$\dot{\phi} = -j\bar{n}_u e^{-\phi j} ,$$

i.e., $\dot{\phi}$ is Lipschitzian. Since $\ddot{r} = -pr$, then $p = -(r, \ddot{r})$ and therefore $\ddot{r} = (r, \ddot{r})r$. Substituting $r = e^{\phi j}$ into this equation yields $\ddot{\phi} \equiv 0$, whence if from the very beginning we take $i = r(t_0)$, then

$$\phi = \omega(t - t_0) + 2\pi k$$

and

$$r = e^{\omega(t-t_0)j} .$$

Let us find ω. Since

$$\dot{r} = \bar{n}_u = \omega j e^{\omega j(t-t_0)} ,$$

then

$$|\omega| = |\bar{n}_u| = \sqrt{1 - |n_u^d|^2} = \sqrt{1 - \frac{|d|^2}{4}} .$$

Thus

$$\rho = \omega^2 ,$$

$$n_u = \frac{d}{2} + \omega j e^{\omega(t-t_0)j} , \qquad n_v = \frac{d}{2} - \omega j e^{\omega(t-t_0)j} ,$$

$$x = x_0 + \int_{t_0}^{t} \frac{n_u}{2} d\tau = x_0 - \frac{i}{2} + \frac{t-t_0}{4} d + \frac{e^{\omega(t-t_0)j}}{2} ,$$

$$y = x - r = x_0 - \frac{i}{2} + \frac{t-t_0}{2} d - \frac{e^{\omega(t-t_0)j}}{2} .$$

The explicit description is obtained.

We see that the regime of ordinary degeneration is a "helical" motion of the segment: its midpoint moves uniformly and rectilinearly, and the segment itself rotates in the plane perpendicular to the motion. The study of standard regimes is complete.

Now we will show that any regime is made up of such regimes. From the described properties of sliding and rotation it is not hard to derive that for any regime given on a segment Δ, if on $\Delta_1 \subset \Delta$ the regime is a sliding (rotation) and Δ_{max} is the segment of the corresponding complete regime, then on $\Delta \cap \Delta_{max}$ the initial regime is a sliding (rotation). The next theorem gives a full classification of regimes.

Theorem 2. Any regime can be extended to a line.

For $\rho \neq \frac{1}{\sqrt{2}}$ there exist the following 6 types of stationary sets on a line:

- 1. successively running complete simple slidings and rotations about alternating endpoints, where the durations of complete slidings and rotations are constant;
- 2. there exists a sequence of segments $\{\Delta_i\}_{i \in I}$ where I is at most countable, the segments are disjoint,*/ $|\Delta_i| \equiv \frac{\pi}{\rho}$, such that on Δ_i the regime is a complete sliding reaching degeneracy, and on $R \setminus \bigcup_{i \in I} \Delta_i$ the regime is an ordinary degenerate one;
- 3. there exist sequences of segments $\{\Delta_i^x\}_{i \in I_x}$, $\{\Delta_i^y\}_{i \in I_y}$, where I_x, I_y are at most countable,

$$I_x \cup I_y \neq \emptyset , \qquad |\Delta_i^x| = |\Delta_i^y| \equiv 2\pi c ,$$

the segments taken from one or from both sequences are disjoint, such that on Δ_i^x, Δ_i^y the regime is a complete rotation with degeneracy reached about the endpoints x, y, while on $R \setminus (\bigcup_{I_x} \Delta_i^x \cup \bigcup_{I_y} \Delta_i^y)$ the regime is ordinary degenerate;
- 4. a sliding;
- 5. a rotation;
- 6. a degenerate regime of ordinary or weighted type.

---

*/ Here and below it is understood that the segments' interiors are disjoint.

Synthesis of the stationary sets can be done under very general assumptions. If $\rho = \frac{1}{\sqrt{2}}$, then the stationary sets are only of types (4.), (5.), (6.) and (3.), with the difference that $|\Delta_i^x| = |\Delta_i^y| \equiv \pi c$. The synthesis problem is solvable in this case under very general assumptions.

For $\rho = 0$ the stationary sets are only of types (1.), (4.), (6.), i.e., extremal motion of the segment is either sequentially running slidings over crossing lines and plane rotations by a constant angle less than π about alternating endpoints, or a helical motion, or a parallel displacement, or a plane motion in which the segment rotates in one direction and the instantaneous centers of rotation lie on it.

Proof. Let a stationary set be given on $\Delta = [a,b]$. We show that it can be extended to $(-\infty,a]$. In fact, if a is degenerate, then there is a degenerate extension. But if a is nondegenerate, the following cases are possible:

- 1. $(r,u)|_a \neq 0$. Then for small $h > 0$ on $[a,a+h]$ our regime is a sliding. We continue it to a complete sliding. If on $[a,a+h]$ the regime is of type 23.1, then the extension to $(-\infty,a]$ has been obtained; if it is of type 23.2, then the left boundary of the extension is degenerate and further extension can be chosen degenerate. If the sliding on $[a,a+h]$ is simple then let α be the left boundary of the segment of complete sliding. Then by 24.1 $|\kappa_\alpha| = |\chi_\alpha|$, $\chi_\alpha s < 0$, and by 27.1 the regime can be extended more to the left as a simple rotation; continuing this rotation to a complete segment up to the point α', we obtain $|\kappa_{\alpha'}| = |\chi_{\alpha'}|$, $s\chi_{\alpha'} > 0$. Again applying 24.1, we extend the regime more to the left of α' as a simple sliding, and so on and so forth. Here the values $|\kappa|$, $|\chi|$, θ are preserved on the junctions of the regimes, so the duration of the regimes of each type do not change. Therefore, the extension to $(-\infty,a]$ is obtained. We construct similarly the extension to $[b,\infty)$;

• 2. $(r,u)|_a = 0$. We take the cases:

2.1 $|n_u(a)| = |n_v(a)| = 1$;

2.2 $|n_u(a)| = |n_v(a)| < 1$;

• 2.1. Here obviously $u(a) = 0$ or $v(a) = 0$ and so by 24.1, $x, y, n_u, n_v|_a$ is the boundary condition of a simple complete sliding. Applying the above procedure we get an extension to $(-\infty,a]$;

• 2.2. here for small $h > 0$ on $[a,a+h]$ the regime is a rotation. It can be extended to $(-\infty,a]$ like the above sliding. The extension to $(-\infty,a]$ is constructed. The extension to $[b,\infty)$ is constructed similarly.

Now let us classify the regimes on R. First we show that if there is a segment Δ on which the regime is a simple sliding or rotation then the regime is of type 1.

Consider the sliding case. Let $[\alpha_0,\beta_0]$ be a complete sliding segment. We show that there is $h > 0$ such that $(r,u) \equiv 0$ on $[\beta_0,\beta_0+h]$. Suppose not. Then for all $h > 0$ there exists $\beta_0 < t^* < \beta_0+h$ such that $(r,u)|_{t^*} \neq 0$. Then in a neighborhood of t^* the regime is a sliding. Let t_ℓ^* be the left boundary of the complete interval of this sliding: $\beta_0 \leq t_\ell^* < t^*$. From 23.3 we conclude that for small h this sliding is simple. Therefore by 24.1, $s\chi_{t_\ell^*} < 0$ and $s\chi_{\beta_0} \leq 0$, which contradicts the fact that on $[\alpha_0, \beta_0]$ the sliding is simple. Thus, for some $h > 0$ on $[\beta_0, \beta_0+h]$ the regime is nondegenerate and $(r,u) = 0$. Then $|u||v| \equiv 0$, and since u, v are continuous, then $u \equiv 0$ or $v \equiv 0$ on $[\beta_0, \beta_0+h]$, i.e., on $[\beta_0, \beta_0+h]$ the regime is a rotation. Let $[\beta_0, \alpha_1]$ be the segment of complete rotation. This rotation is simple since β_0 is nondegenerate. Let us show that there exists $h > 0$ such that $(r,u) \neq 0$ on $[\alpha_1, \alpha_1+h]$. Suppose not. Then the following cases are possible:

- 1. for some $h > 0$ one has $(r,u) \equiv 0$ on $[\alpha_1, \alpha_1+h]$. Then on $[\alpha_1, \alpha_1+h]$ our regime is a rotation, and a simple one, since

$$2s(\theta+s) + \frac{2A|s|}{c}\Big|_{\alpha_1} > 0 .$$

But then α_1 is simultaneously a right and left boundary condition of a simple rotation, which contradicts 27.1;

- 2. for any $h > 0$ on $(\alpha_1, \alpha_1+h]$ there are points where $(r,u) = 0$ and $(r,u) \neq 0$. But then there is a segment $\Delta \subset [\alpha_1, \alpha_1+h]$ on which the regime is a simple complete sliding. By 26.4, $\kappa_{\beta_0} = -\kappa_{\alpha_1}$, $\chi_{\beta_0} = -\chi_{\alpha_1}$, $\theta_{\beta_0} = \theta_{\alpha_1}$; and since the length of an interval of a complete simple sliding depends continuously on $|\kappa|$, $|\chi|$, θ, then for small h, $|\Delta| \geq \frac{\beta_0+\alpha_0}{2}$, which contradicts $|\Delta| \leq h$ and the arbitrary smallness.

Thus there is an $h > 0$ such that on $[\alpha_1, \alpha_1+h]$ our regime is a simple sliding.

Let β_1 be the right boundary of a complete sliding. Continuing the procedure we construct the sequence

$$\cdots < \alpha_{-1} < \beta_{-1} < \alpha_0 < \beta_0 < \alpha_1 < \beta_1 < \cdots$$

such that on $[\alpha_i,\beta_i]$ the regime is a complete simple sliding and on $[\beta_i,\alpha_{i+1}]$ a complete simple rotation. By 23.4, 26.4, the $|\kappa|$, $|\chi|$, θ are preserved at α_i, β_i and $\kappa\chi|_{\alpha_i} = -\kappa\chi|_{\beta_i}$, $\kappa\chi|_{\alpha_{i+1}} = \kappa\chi|_{\beta_i}$. Whence $\beta_i - \alpha_i$, $\alpha_{i+1} - \beta_i \equiv$ const. and the rotations on $[\beta_{i-1},\alpha_i]$, $[\beta_i,\alpha_{i+1}]$ go about different endpoints. Thus our regime is of type 1.

Consider next stationary sets for $\rho \neq \frac{1}{\sqrt{2}}$, i.e., for $s \neq 0$. The following two cases are possible:

- a. there are no degeneracy points on R, and
- b. there are degeneracy points on R.

• a. In this case there is a segment Δ on which the regime is a complete sliding or rotation not reaching degeneracy. By (23), (26) this means that either $\Delta = R$ or the regime is simple on Δ. If $\Delta = R$, then our regime is of type 4 or type 5, otherwise the regime is of type 1, as shown above.

• b. In this case, if there are no nondegenerate points on R, the regime is degenerate, i.e., is of type 6. For nondegenerate points, consider the sets $Q = \{t \mid \text{is nondegenerate}\}$, $F = \{t \in Q \mid (r,u) \neq 0\}$. On intervals of the open set F the regime is obviously a complete sliding reaching degeneracy. Their length, according to 23.2, is $\frac{\pi}{\rho}$. Since the boundary of F is discrete then on intervals of the open set $Q \setminus \bar{F}$ the regime is a complete rotation reaching degeneracy. The length of these intervals, by 26.2, is $2\pi c$. Let the components of $\bar{F}$, $\overline{Q\setminus\bar{F}}$ be denoted by Δ_i, Δ_i^x, Δ_i^y, respectively. Then the segments of this family are mutually disjoint,

$$|\Delta_i| \equiv \frac{\pi}{\rho}, \qquad |\Delta_i^x| = |\Delta_j^y| \equiv 2\pi c ,$$

while on the set

$$R \setminus (\bigcup_i \Delta_i \cup \bigcup_i \Delta_i^x \cup \bigcup_i \Delta_i^y)$$

the regime is degenerate.

Let us show that it is impossible that $F \neq \emptyset$, $Q \setminus \bar{F} \neq \emptyset$. Suppose not. Let (α_1, β_1), (α_2, β_2) be components of these sets such that on (β_1, α_2) there are no points of Q. By 24.2, 26.2 one has $s(\theta_{\beta_1} - s) > 0$, $s(\theta_{\alpha_2} - s) < 0$. These inequalities are incompatible in θ and hence $\beta_1 \neq \alpha_2$. But then on $[\beta_1, \alpha_2]$ the regime is degenerate. Therefore, $\theta \equiv \text{const.}$ on $[\beta_1, \alpha_2]$, and we again arrive at a contradiction due to the incompatibility of the inequalities in θ.

Thus, if $F = \emptyset$ our regime is of type 3, but if $F \neq \emptyset$ our regime is of type 2.

The construction of the synthesis is obvious. For $\rho = 0$, i.e., in the spatial case, according to (23), (26), only (1.), (4.), (6.) are realized. The case when $\rho = \frac{1}{\sqrt{2}}$ is considered in exactly the same way. //

Starting from the derived classification of extremals, it is not hard to construct an atlas of optimal motions of segments on the plane, making it possible to find, for any given segments, their optimal motion by simple geometric constructions.

We give some examples.

Consider the case in Figure 1, where the optimal motion is of rotation of $\vec{AB}$ to position $\vec{AC}$, then comes a rotation about C to position $\overrightarrow{A'C}$, then a rotation about A' to position $\overrightarrow{A'B'}$ It is interesting to note that in the considered case the minimal sum of the lengths of the paths swept out by the endpoints of the

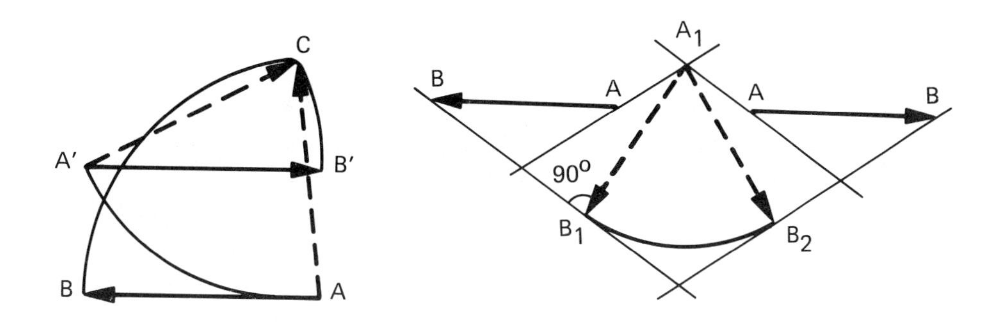

Figure 1 Figure 2

segment under motion is π and does not depend on the distance between the segments.

Consider the optimal motion of the segments laid out as in Figure 2. Here the segment slides from $\overrightarrow{AB}$ to $\overrightarrow{A_1B_1}$, then it rotates about A_1 to position $\overrightarrow{A_1B_2}$ and slides to position $\overrightarrow{A'B'}$.

Let us go to Figure 3. Here the optimal motion is a rotation of $\overrightarrow{AB}$ about A to position $\overrightarrow{AC}$, then a sliding to position $\overrightarrow{C'B'}$ and a rotation about B' to position $\overrightarrow{A'B'}$.

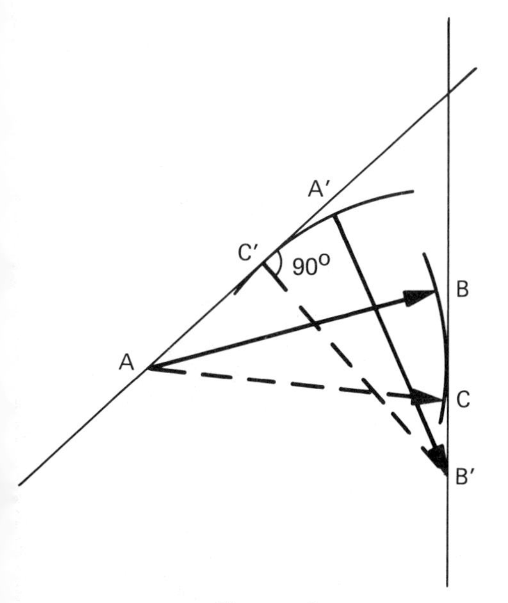

Figure 3

• • • • •

CHAPTER 3

OPTIMAL MOTION OF SEGMENTS ON THE CIRCULAR CYLINDER

Now we examine Ulam's problem on the cylinder $S^1 \times R$ in the space E^3. An essential role in the solution is played by the fact that on a cylinder the operation of parallel displacement of a vector does not depend on the path of the displacement and in displacement from x to y, $\bar{r}_1$ goes over to $\bar{r}_2$, where $r = x - y$.*/ Owing to this property, as has been observed (see Section 1.5), after displacement of $n_u(t)$, $n_v(t)$ to any common point we have $n_u(t) + n_v(t) \equiv \text{const.} = d$.

An equation of the cylinder is $g(x) = |\bar{x}|^2 - R^2 = 0$, where $\bar{x}$ is the projection of x onto the base plane. Take also an orthonormal basis i, j, k, where k is directed along the axis of the cylinder, and let the projection of x in the direction k be denoted $x_\perp$.

The stationary set equations (20) take the form:

---

*/ Recall that $\bar{r}_1$, $\bar{r}_2$ denote the projections of the vector $r = x-y$ onto the tangent planes $V(x)$, $V(y)$, respectively.

$$\dot{n}_u = -pr - 2p_1\bar{x} , \qquad \dot{n}_v = pr - 2p_2\bar{y} . \tag{29}$$

Let us turn to studying the typical regimes. For convenience in describing them, introduce on the plane i, j the complex algebra, i.e., we will identify the vector $xi + yj$ with the complex number $x + yj$.

§ 1. THE SLIDING REGIME

We will let ϕ denote the angle between $\bar{x}$ and $\bar{y}$. The geodesics of the cylinder on which the endpoints of the segment slide, are helical curves in this regime. Let us write them in the form:

$$\begin{aligned} q_1(\alpha) &= (x_\perp^0 + \alpha V_x)k + \bar{x}_0 e^{\alpha\omega_x j} ; \\ q_2(\beta) &= (y_\perp^0 + \beta V_y)k + \bar{y}_0 e^{\beta\omega_y j} , \end{aligned} \tag{30}$$

where α, β are the lengths of the arcs and therefore

$$V_x^2 + R^2\omega_x^2 = V_y^2 + R^2\omega_y^2 = 1 .$$

The support functions n_u, n_v have the form:

$$n_u = V_x k + \bar{x}_0 j \omega_x e^{\alpha\omega_x j} ,$$

$$n_v = V_y k + \bar{y}_0 j \omega_y e^{\beta\omega_y j}$$

and

$$\begin{aligned} \sigma &= (n_u, r) = V_x r_\perp - \omega_x R^2 \sin\phi , \\ \gamma &= (n_v, r) = V_y r_\perp - \omega_y R^2 \sin\phi . \end{aligned} \tag{31}$$

It is more convenient for us to change from the variables α, β to the "cylindrical" ones $\phi = (\widehat{\bar{x},\bar{y}})$, $r_\perp$. From (30) it follows that

$$\begin{aligned} r_\perp - r_\perp^0 &= \alpha V_x - \beta V_y \ , \\ \phi - \phi_0 &= \beta\omega_y - \alpha\omega_x \ . \end{aligned} \tag{32}$$

Let δ denote

$$\begin{vmatrix} V_x & \omega_x \\ V_y & \omega_y \end{vmatrix} \ .$$

The equations of the general sliding representation

$$\dot{\alpha} = \frac{\gamma}{\sigma+\gamma} \ , \qquad \dot{\beta} = \frac{\sigma}{\sigma+\gamma} \ ,$$

in combination with (31), (32), give:

$$\dot{r}_\perp = -\delta R^2 \frac{\sin\phi}{\sigma+\gamma} \ , \qquad \dot{\phi} = \delta \frac{r_\perp}{\sigma+\gamma} \ . \tag{33}$$

Depending on δ, two cases are possible.

- 1. $\delta = 0$. In this case $(V_x, \omega_x) \| (V_y, \omega_y)$ and since

$$V_x^2 + R^2\omega_x^2 = V_y^2 + R^2\omega_y^2 = 1 \ ,$$

then

$$(V_x, \omega_x) = \pm(V_y, \omega_y) \ .$$

But the sign "-" is inadmissible since then $\sigma = -\gamma$ and $\sigma\gamma < 0$, which contradicts the fact that under the sliding, $\sigma\gamma > 0$ in $\operatorname{int}\Delta$. Thus, $V_x = V_y$, $\omega_x = \omega_y$. Consequently

$$\sigma = \gamma \ ,$$

$$\dot{\alpha} = \frac{\gamma}{\sigma+\gamma} = \frac{1}{2} = \dot{\beta} \ ,$$

$$\alpha = \beta = \frac{t}{2} \ .$$

Our regime thus has the form:

$$x = \left(x_{\perp}^{0} + \frac{t}{2} V_x\right) k + \bar{x}_0 \, e^{\frac{t}{2}\omega_x j} ,$$

$$y = \left(y_{\perp}^{0} + \frac{t}{2} V_y\right) k + \bar{y}_0 \, e^{\frac{t}{2}\omega_y j} ,$$

and $(r_0, n_u^0) \neq 0$.

This is a parallel displacement of the segment on the cylinder in a direction nonorthogonal to the segment.

• 2. $\delta \neq 0$. Note first of all that in this case sliding cannot reach a position of degeneration. Otherwise at such a point one has $\sigma = \gamma = 0$ or

$$V_x r_{\perp} - \omega_x R^2 \sin\phi = 0 , \qquad V_y r_{\perp} - \omega_y R^2 \sin\phi = 0 .$$

Whence since $\delta \neq 0$ then $r_{\perp} = 0 = \sin\phi$. Since

$$\bar{r}_1 = -\sin\phi \cdot j\bar{x} + r_{\perp} ,$$

then $\bar{r}_1 = 0$, which contradicts the assumption $|\bar{r}_1||\bar{r}_2| > 0$ on the studied trajectories.

System (33) has first integral $r_{\perp}^2 + 2R^2(1 - \cos\phi)$ (this is the length of the segment). We sketch the level curves $r_{\perp}^2 + 2R^2(1 - \cos\phi) = 1$:

• a. $R > \frac{1}{2}$ (Figure 4)*/;

• b. $R = \frac{1}{2}$ (Figure 5);

• c. $R < \frac{1}{2}$ (Figure 6).

Let us study the properties of a complete sliding. The inter-

*/ On Figure 4, $\alpha = \arccos\left(1 - \frac{1}{2R^2}\right)$.

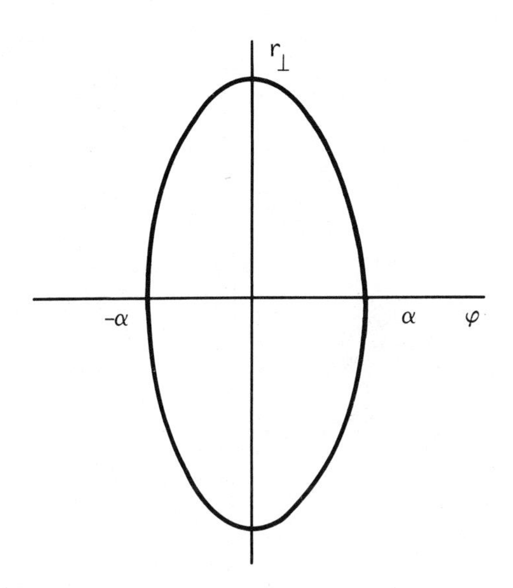

Figure 4

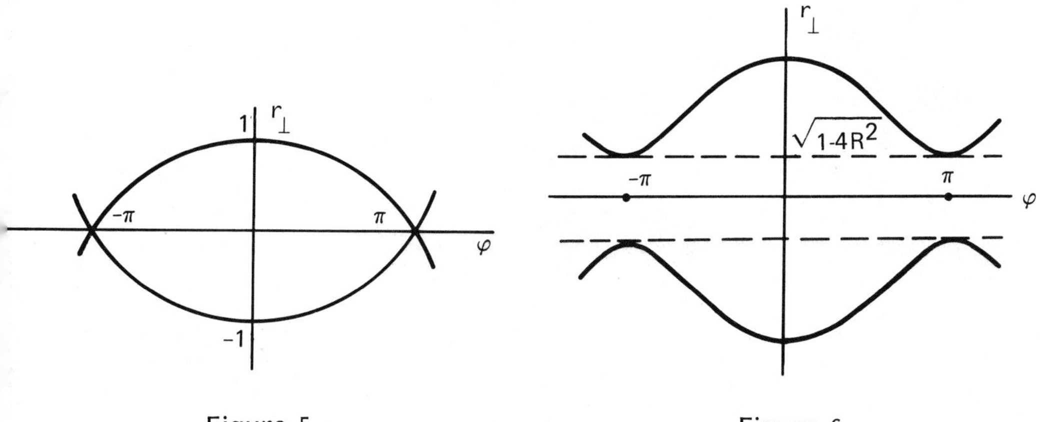

Figure 5

Figure 6

val of complete sliding [a,b] is determined by the condition $\sigma\gamma > 0$ in (a,b). Depending on the mutual arrangement of the curves

$$\ell = \{r_\perp^2 + 2R^2(1 - \cos\phi) = 1\} ,$$

$$\ell_1 = \{\delta = V_x r_\perp - R^2\omega_x \sin\phi = 0\} ,$$

$$\ell_2 = \{\gamma = V_y r_\perp - R^2\omega_y \sin\phi = 0\} ,$$

the following situations are possible on the plane $\phi, r_\perp$:

- a. $\ell \cap (\ell_1 \cup \ell_2) = \emptyset$. Here $\sigma\gamma > 0$ for any t and "perpetual" sliding occurs;
- b. $\ell \cap \ell_1 = \emptyset$, $\ell_1 \cap \ell_2 \neq \emptyset$ or $\ell \cap \ell_1 \neq \emptyset$, $\ell \cap \ell_2 = \emptyset$. Here the component $\sigma\gamma > 0$ is finite, and in the first case $\gamma_a = \gamma_b = 0$ and in the second $\sigma_a = \sigma_b = 0$;
- c. $\ell \cap \ell_1 \neq \emptyset$, $\ell \cap \ell_2 \neq \emptyset$. Here $[a,b]$ is also finite. For the behavior of σ, γ on the boundary for $R < \frac{1}{2}$ any variants are possible: while for $R \geq \frac{1}{2}$, if $\sigma_a = 0$ then $\gamma_b = 0$, and vice versa.

Note that for $R \geq \frac{1}{2}$ only case (c.) is realized, while for $R < \frac{1}{2}$, all three are realized.

Using

$$\xi_u = (n_u, j\bar{x}) \quad , \qquad \xi_v = (n_v, j\bar{y}) \quad ,$$

we can represent the conditions

$$\ell \cap \ell_1 = \emptyset \quad , \qquad \ell \cap \ell_2 = \emptyset$$

in the form

$$\xi_u > \psi \quad , \qquad \xi_v > \psi \quad ,$$

where

$$\psi = \arctan \sqrt{\frac{1 - 2R^2 - \sqrt{1-4R^2}}{2R^2}} \quad ,$$

and write the cases chosen as

- a. $\xi_u, \xi_v > \psi$;
- b. $\max(\xi_u, \xi_v) > \psi$, $\min(\xi_u, \xi_v) \leq \psi$;
- c. $\xi_u, \xi_v \leq \psi$.

If we know d, then n_u, n_v (all carried to one point) are determined to within a permutation. Hence d determines the pairs V_x, ω_x and V_y, ω_y to within a permutation. From this it follows that $b-a$ is uniquely determined for $R \geq \frac{1}{2}$ and can take on at most two values for $R < \frac{1}{2}$.

§ 2. THE ROTATING REGIME

Let us examine rotation about the endpoint "x". We immediately deduce its representation from the general one (see Section 1.6.b):

$$p = \frac{r_\perp^2 \cos\phi + R^2 \sin^2\phi}{|\bar{r}_1|^2} ,$$

$$p_1 = \frac{p(\cos\phi - 1)}{2} , \tag{35}$$

$$p_2 = \frac{r_\perp^4 + r_\perp^2 + R^2(r_\perp^2 - 1)\sin\phi}{4R^2|\bar{r}_1|^2} ,$$

where

$$\dot{n}_u = -p\bar{r}_1 , \qquad \dot{n}_v = pr - 2p_2\bar{y}_2 ,$$

$$\dot{y} = n_v ,$$

$$|n_v^0| = |r_0| = 1 ,$$

$$(n_u^0, \bar{x}_0) = (n_v^0, \bar{y}^0) = (n_v^0, r_0) = 0$$

and in $\operatorname{int}\Delta$ there are no degenerate points, $|n_u| \leq 1$.

Let us show that this regime cannot reach degenerate positions. Suppose not. Then at a degenerate point

$$|n_u| = |n_v| = 1 , \qquad (n_u, r) = (n_v, r) = 0 .$$

Hence, since $\bar{r}_1 = -\sin\phi j\bar{x} + r_\perp k$ and n_u, n_v are tangent to the cylinder, then

$$n_u = \varepsilon_u \frac{r_\perp \frac{j\bar{x}}{R} + R\sin\phi \cdot k}{|\bar{r}_1|} ,$$

$$n_v = \varepsilon_v \frac{r_\perp \frac{j\bar{y}}{R} + R\sin\phi \cdot k}{|\bar{r}_2|} ,$$

where $|\varepsilon_u| = |\varepsilon_v| = 1$. The vector d carried to i is

$$(\varepsilon_u + \varepsilon_v) \frac{r_\perp j + R\sin\phi \cdot k}{|\bar{r}_1|} .$$

The following cases are possible:

• 1. $\varepsilon_u \varepsilon_v = -1$. Then $d = 0$ and therefore the regime is purely degenerate. Contradiction.

• 2. $\varepsilon_u \varepsilon_v = 1$. Then $|d| = 2$ and therefore $|n_u| = |n_v| = 1$. From (35) we conclude that

$$\frac{d}{dt} |n_u|^2 = -2p\sigma \equiv 0 .$$

Let us sort out the cases:

• a. There exists $t \in \text{int}\,\Delta$, where $p(t) \neq 0$. But then $\sigma(t) = 0$, i.e., t is degenerate. Contradiction.

• b. $p \equiv 0$ in $\text{int}\,\Delta$. Then, since $p = 0$ implies

$$\cos\phi = \frac{2R^2 - 1 + \sqrt{1-4R^2}}{2R^2} ,$$

one has $\phi \equiv$ const., which is impossible under a rotation of the segment.

Thus a degenerate position cannot be attained.

Next we shall examine the complete rotation. In this problem it is advisable to change the definition of a complete rotation. By a complete rotation we will mean an extension of the solution of the system (35) to a maximal interval interior to which $|n_u| < 1$. Under this definition, obviously, if some regime contains a section of rotation and Δ_{max} is the interval of the corresponding complete rotation, then on $\Delta \cap \Delta_{max}$ our regime is a rotation. For convenience, we take n_u, n_v d to the point i. Then $|n_u| < 1$ can be replaced by $|d - n_v| < 1$. Since n_v moves in the tangent plane j, k and $n_v \perp r$, then n_v is determined to within the sign by $\bar{r}_1$. Hence, knowing the hodograph of $\bar{r}_1$, we can also easily determine the course of motion of n_v to within direction on the unit circumference.

For definiteness we assume that $\bar{r}_1$, n_v are oriented like j, k and that n_v moves in the positive direction. Then

- a. For $R > \frac{1}{2}$, $\bar{r}_1$ moves over the closed curve

$$\bar{r}_1 = -\sin \phi R j + \varepsilon \sqrt{1 + 2R^2(\cos \phi - 1)}\ k ,$$

where $\varepsilon = \pm 1$, and n_v describes a complete circle (see Figure 7).

- b. For $R \leq \frac{1}{2}$, $\bar{r}_1$ circulates along the curve (see Figure 8).

$$\bar{r}_1 = -\sin \phi R j + \varepsilon \sqrt{1 + 2R^2(\cos \phi - 1)}\ k ,$$

where $\varepsilon = \pm 1$, and n_v oscillates on the arcs $\smile{AB}$ or $\smile{A'B'}$, and */

*/ The extreme positions of n_v are reached for

$$\cos \phi = \frac{2R^2 - 1 + \sqrt{1 - 4R^2}}{2R^2} .$$

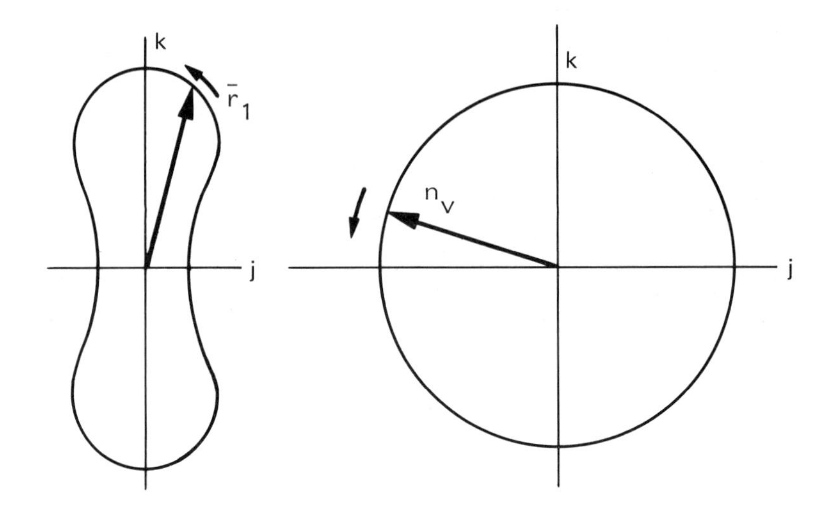

Figure 7

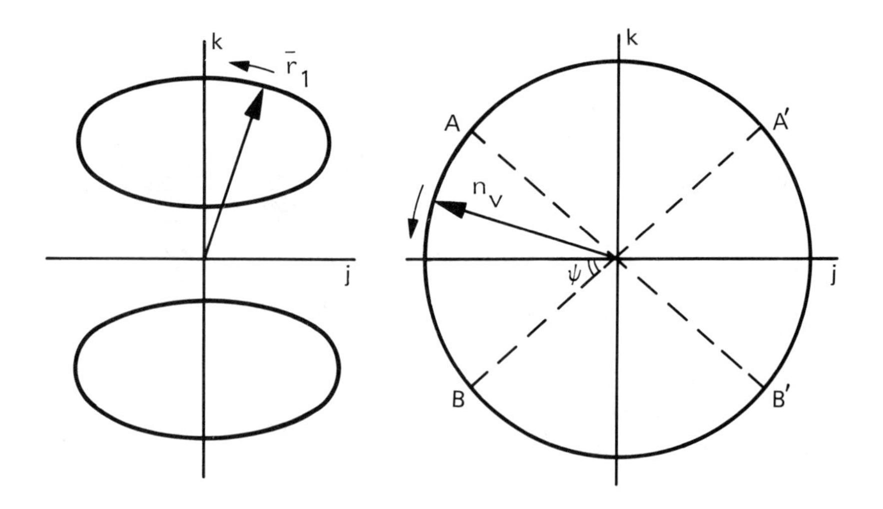

Figure 8

$$\psi = \arctan \sqrt{\frac{1 - 2R^2 - \sqrt{1-4R^2}}{2R^2}} .$$

Under rotation $0 < |d| < 2$. Depending on d, R, the condition $|d - n_v| < 1$ distinguishes the following types of complete rotation:

• 1. for $R > \frac{1}{2}$ the segment $[a,b]$ is finite and $n_u, n_v|_a = n_v, n_u|_b$;

• 2. for $R \leq \frac{1}{2}$ the following cases are possible:

• 2.1. $|d-A| < 1$, $|d-B| < 1$. Then "perpetual" rotation occurs.

• 2.2 $|d-A| < 1$, $|d-B| \geq 1$, or conversely. Then $[a,b]$ is finite and $n_u, n_v|_a = n_u, n_v|_b$;

• 2.3. $|d-A| \geq 1$, $|d-B| \geq 1$. Then $[a,b]$ is finite and $n_u, n_v|_a = n_v, n_u|_b$.

Let the segment $[a,b]$ be finite. Then if d is given, since $|n_u| = 1$, $|n_v| = 1$, n_u, n_v are determined to within a permutation. From this it follows that for $R > \frac{1}{2}$, $b - a$ takes on a unique value, and for $R \leq \frac{1}{2}$ at most two values (for fixed d).

Now it is advisable to study the possibility of combining complete slidings and rotations.

Obviously a complete rotation can be extended by a rotation only for $R \leq \frac{1}{2}$ and $|d-A| = |d-B| = 1$, i.e., for

$$d = \frac{2\sqrt{2}\, R}{\sqrt{1 - \sqrt{1-4R^2}}}\, j .$$

Now let us study the possibility of extending a complete

sliding by a sliding. As we know the segment of a complete sliding $[a,b]$ is a maximal segment interior to which the solution of the system

$$\dot{r}_\perp = -\delta R^2 \frac{\sin\phi}{\sigma+\gamma}, \qquad \dot{\phi} = \delta \frac{r_\perp}{\sigma+\gamma}$$

satisfies $\sigma\gamma > 0$.

For definiteness, consider the point b, where $\gamma_b = 0$, $\sigma_b \neq 0$. For the existence of an extension of the regime to the right by a sliding it is obviously necessary that $\frac{d}{dt}\sigma\gamma_{|b} = 0$. As a result of the calculation we obtain the expression

$$\frac{d}{dt}\sigma\gamma_{|b} = -\sigma(V_y^2 + R^2\omega_y^2\cos\phi) .$$

Whence

$$\mathrm{tg}\,\xi_v = \sqrt{\frac{1 - 2R^2 - \sqrt{1-4R^2}}{2R^2}}$$

and

$$R \leq \frac{1}{2},$$

where $\xi_v = (n_v, j)$.

Let us show that fulfillment of this condition is sufficient for the existence of an extension of a sliding. Indeed,

$$\frac{d^2}{dt^2}\sigma\gamma_{|b} = \sigma\omega_y^3 R^2 \sin\phi .$$

Let us show that $\omega_y \sin\phi \neq 0$. The relations

$$V_y^2 + R^2\omega_y^2 = 1 ,$$

$$\gamma = V_y r_\perp - \sin\phi R^2 \omega_y = 0 ,$$

$$V_y^2 + R^2\omega_y^2 \cos\phi = 0$$

yield $\omega_y \neq 0$. But if $\sin\phi = 0$, then $\phi = 0$ or $\phi = \pi$. The case $\phi = 0$ is impossible since then

$$V_y^2 + R^2\omega_y^2 \cos\phi = V_y^2 + R^2\omega_y^2 = 0 ,$$

a contradiction. The case $\phi = \pi$ leads to $\bar{r}_1 = 0$, which is also impossible for $R \leq \frac{1}{2}$. Thus $\frac{d^2}{dt^2}\sigma\gamma_{|b} \neq 0$ and since $\sigma\gamma > 0$ on (a,b), $\frac{d}{dt}\sigma\gamma_{|b} = 0$, then $\frac{d^2}{dt^2}\sigma\gamma_{|b} > 0$. Consequently, $\sigma\gamma > 0$ for $t > b$ and an extension of sliding type exists.

It is shown similarly that any complete sliding is extendable by a rotation, and conversely.

§ 3. THE DEGENERATE REGIME

This regime is characterized by the fact that

$$|n_u| = |n_v| \equiv 1 ,$$

$$(n_u, r) = (n_v, r) \equiv 0 .$$

Since n_u, n_v are tangent to the cylinder, then

$$n_u = \varepsilon_u \frac{r_\perp \frac{j\bar{x}}{R} + \sin\phi \cdot Rk}{|\bar{r}_1|} ,$$

$$n_v = \varepsilon_v \frac{r_\perp \frac{j\bar{y}}{R} + \sin\phi \cdot Rk}{|\bar{r}_1|} ,$$

where $|\varepsilon_u| = |\varepsilon_v| = 1$.

The following cases are possible:

• 1. $\varepsilon_u \varepsilon_v = 1$. Here $|d| = 2$. Hence moving n_u, n_v to i, we have $n_u = \frac{d}{2}$, $n_v = \frac{d}{2}$. From the expressions for n_u, n_v we conclude that $\phi \equiv \text{const}$. The representation of the degenerate regime gives us

$$p = \frac{(\alpha-\beta)(\cos\phi + R^2(\cos\phi-1)^2)}{(1 - R^2(1 - \cos\phi)^2)^2} ,$$

where α, β are measurable weights.

On the other hand, condition (21) here takes on the form:

$$(\alpha-\beta)(\cos\phi + R^2(\cos\phi-1)^2) \equiv 0 .$$

Thus $p \equiv 0$, whence from (29), $\dot{n}_u = -2p_1\bar{x}$, $\dot{n}_v = -2p_2\bar{y}$. This means that the endpoints of the segment move on helical curves; since $\dot{n}_u = \frac{d}{2}$, $\dot{n}_v = \frac{d}{2}$, these curves are parallel. Let μ, ν denote the lengths of the arcs. We can write:

$$x = (x_\perp^0 + \mu V)k + \bar{x}_0 e^{\mu\omega j} ,$$

$$y = (y_\perp^0 + V\nu)k + \bar{y}_0 e^{\nu\omega j} ,$$

where $V^2 + R^2\omega^2 = 1$. Finally, since

$$\phi = (-\mu + \nu)\omega + \phi_0$$

and

$$|\dot{x}| + |\dot{y}| = 1 ,$$

our regime is a parallel displacement of the segment on the cylinder in direction orthogonal to the segment.

• 2. $\varepsilon_u \varepsilon_v = -1$. Here $d = 0$. From the general representation of degenerate regimes we immediately deduce a description of this regime by the system:

$$\dot{n}_u = -pr - 2p_1\bar{x} \quad ;$$

$$\dot{n}_v = pr - 2p_2\bar{y} \quad ;$$

$$\dot{x} = \alpha n_u \ , \qquad \dot{y} = \beta n_v \ ,$$

where

$$p = \frac{\cos\phi + R^2(\cos\phi - 1)^2}{(1 - R^2(1-\cos\phi)^2)^2} \ ,$$

$$p_1 = \frac{\alpha|\bar{n}_u|^2 + pR^2(\cos\phi - 1)}{2R^2} \ ,$$

$$p_2 = \frac{\beta|n_v|^2 + pR^2(\cos\phi - 1)}{2R^2}$$

and α, β are any measurable weights (i.e., nonnegative measurable functions with $\alpha + \beta = 1$). At some $t_0 \in \Delta$ the set $x, y, n_u, n_v|_{t_0}$ must be degenerate, and $n_u(t_0)$, $n_v(t_0)$ are opposites to within a parallel shift. Just as before, we call this the "weight" regime.

§ 4. DESCRIPTION OF EXTREMALS

It is easy to deduce a complete classification of extremals.

Theorem 3. Every regime can be extended to a line (in the parameter t). Depending on d and R, the following types of regimes exist:

- a. $d = 0$, the "weight" degenerate regime described above;
- b. $|d| = 2$, a parallel displacement of the segment on a cylinder, i.e., a sliding of the endpoints of the segment along

parallel helical curves;

- c. $0 < |d| < 2$, the regime consists of a succession of combinant complete slidings along nonparallel helical curves and complete rotations.

For $R > \frac{1}{2}$ the slidings, rotations and segment endpoints about which the rotations occur alternate, and the duration of complete slidings and rotations are constants.

For $R \leq \frac{1}{2}$ either the whole regime is a "perpetual" sliding or rotation, or its component slidings and rotations have, correspondingly, durations taking on at most two values. The sequence of combinant regimes depends on the mutual positioning of the vectors */

$$n_u,\ n_v,\ \tilde{d} = \frac{2\sqrt{2}\ R}{\sqrt{1 - \sqrt{1-4R^2}}} \quad ,$$

where n_u, n_v are taken at a point t such that $|n_u| = |n_v| = 1$.

Let $\tilde{d} = c_1^1 + c_2^1$, where $|c_1^1| = |c_2^1| = 1$. Set $c_1^{-1} = -c_1^1$, $c_2^{-1} = -c_1^1$. Then:

- 1. if

$$\{n_u, n_v\} \cap \bigcup_{\nu=\pm 1} \{c_1^\nu, c_2^\nu\} = \emptyset \quad ,$$

then the regime is either a perpetual sliding or consists of alternating slidings and rotations;

- 2. if

$$\{n_u, n_v\} \cap \{c_1^\nu, c_1^\nu\} \neq 0 \quad ,$$

$$\{n_u, n_v\} \neq \{c_2, c_2\}$$

*/ We assume that n_u, n_v are moved by a parallel shift to i.

for $\nu = 1$ or $\nu = -1$, then complete slidings can combine, but rotations cannot;

• 3. if $\{n_u, n_v\} = \{c_1^\nu, c_2^\nu\}$ for $\nu = 1$ or $\nu = -1$, then both complete slidings and rotations can combine.

Proof. The proof is carried out here exactly as in solving the spatial-spherical problem with the exception that the lower estimate for the durations of the sections of complete slidings and rotations is deduced from the constancy of d. We will not repeat it.

Thus the supply of extremals depends on the radius of the cylinder. For $R > \frac{1}{2}$ the picture is the same as in the planar Ulam problem, while for $R \leq \frac{1}{2}$ some new results appear. For example, there are possible "perpetual" rotations and slidings that are not a parallel shift, or a succession or slidings and rotations in which the fixed endpoints do not alternate.

Let us give a few examples.

• 1. Sliding of the endpoints of the segment along parallel helical curves that degenerate into the circles $(R < \frac{1}{2})$

$$x(t) = \sqrt{1 - 4R^2} + R\, e^{\frac{t}{2}j} ,$$

$$y(t) = -R\, e^{\frac{t}{2}j}$$

(see Figure 9).

• 2. "Perpetual" rotation on the cylinder for $R < \frac{1}{2}$ (see Figure 10).

• 3. The "weighted" degenerate regime with weight functions $\alpha = \rho \equiv \frac{1}{2}$. In this regime the midpoint of the segment lies on a line perpendicular to the axis of the cylinder (see Figure 11).

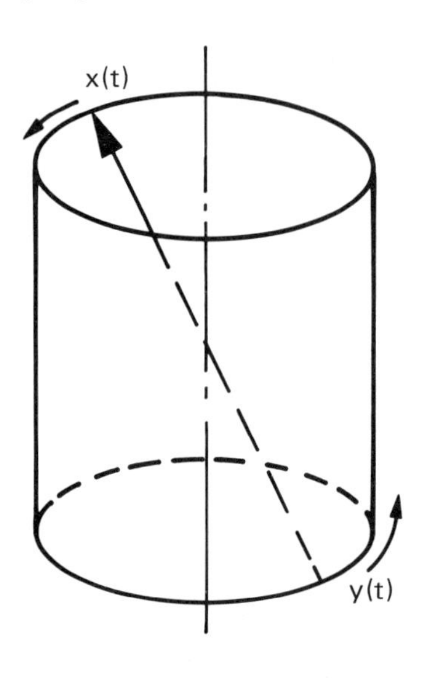

Figure 9

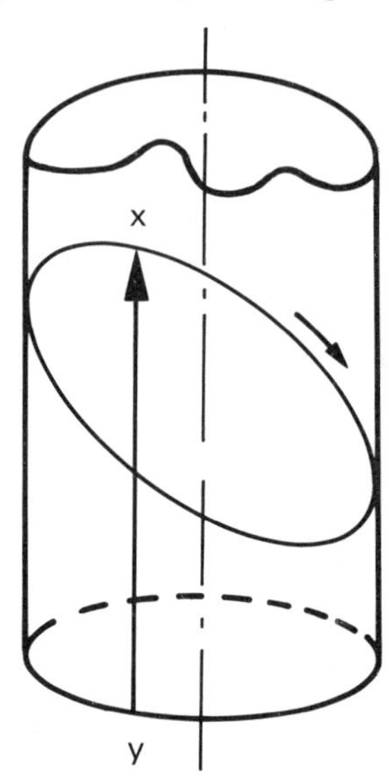

Figure 10

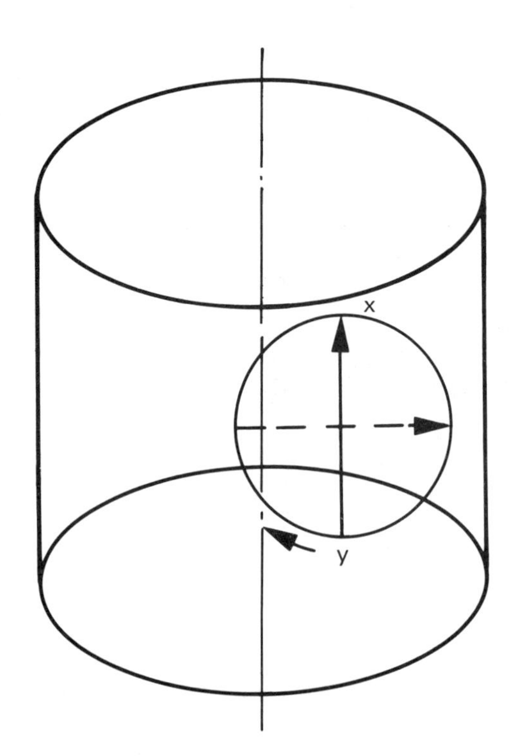

Figure 11

CHAPTER 4

OPTIMAL MOTION OF SEGMENTS ON THE ANISOTROPIC PLANE

In conclusion we examine a variation of Ulam's problem: we will use different Euclidean metrics to measure the length of the segment and that of the paths swept out by its endpoints.

The presence of the second metric (the one used to measure path length) does not change anything in the variational analysis of the problem, but generally speaking, the properties of the extremals get very complex. We have been fully successful in describing only the planar extremals of this problem. It has turned out that the extremals of the planar problem generally do not depend on the second metric (to within a re-parametrization). This circumstance is special to the plane: in higher dimensions there is no such property.

We will write the scalar product defining the second metric in the form $(x,y)_A = (x,Ay)$, where A is a positive definite operator. We let $|\cdot|_A$ denote the norm induces by the scalar product $(\cdot,\cdot)_A$. The stationary set equations in this problem take the form

- 1. $\dot{n}_u = -pr$, $\dot{n}_v = pr$;
- 2. $|\dot{x}|_A + |\dot{y}|_A = 1$, $|x-y| = 1$; (36)

n_u, n_v are supporting to $|\cdot|_A$ at the points $\dot{x}$, $\dot{y}$.

Recall that "n_u is supporting to $|\cdot|_A$" for $z \neq 0$ means that $n_u = \frac{Az}{|z|_A}$, while for $z = 0$, $|n_u|_{A^{-1}} \leq 1$.

The degeneracy conditions take the form:

$$|n_u|_{A^{-1}} = |n_v|_{A^{-1}} = 1 ,$$

$$(n_u, r)_{A^{-1}} = (n_v, r)_{A^{-1}} = 0 .$$

From (36) we deduce that $\dot{n}_u + \dot{n}_v = 0$. Hence $n_u + n_v = d \equiv \text{const}$. This circumstance makes it possible (just as in the cylindrical case) to easily solve the problem for $n = 2$.

Before analyzing the regimes, we note that from the supportingness of n_u, n_v it follows that

$$\dot{x} = |\dot{x}|_A A^{-1} n_u , \qquad \dot{y} = |\dot{y}|_A A^{-1} n_v . \tag{37}$$

First consider the regime on Δ containing a degenerate point $t^* \in \Delta$. At this point

$$|n_u|_{A^{-1}} = |n_v|_{A^{-1}} = 1 ,$$

$$(n_u, r)_{A^{-1}} = (n_v, r)_{A^{-1}} = 0 .$$

Since $r \neq 0$, then for $n = 2$ one has $n_u = \varepsilon n_v$, where $\varepsilon = \pm 1$. If $\varepsilon = 1$, then $|d|_{A^{-1}} = 2$. Hence, since

$$|n_u|_{A^{-1}}, \quad |n_v|_{A^{-1}} \leq 1 ,$$

$$n_u + n_v \equiv d ,$$

then $n_u = n_v \equiv \frac{d}{2}$. From equations (37) we now deduce that

$$x = x_0 + \int_{t_0}^{t} |\dot{x}| \, d\tau \, \frac{A^{-1}d}{2} ,$$

$$y = y_0 + \int_{t_0}^{t} |\dot{y}| \, d\tau \, \frac{A^{-1}d}{2} .$$

Finally, since $|r| \equiv 1$ then

$$1 = |x-y|^2 = 1 + \left(\int_{t_0}^{t} (|\dot{x}| - |\dot{y}|) \, d\tau \, A^{-1}d, \; r_0\right)$$
$$+ \int_{t_0}^{t} (|\dot{x}| - |\dot{y}|) \, d\tau \left|\frac{A^{-1}d}{2}\right|^2 .$$

Whence

$$\int_{t_0}^{t} |\dot{x}| \, d\tau = \int_{t_0}^{t} |\dot{y}| \, d\tau = \frac{t-t_0}{2} ,$$

and

$$x = x_0 + \frac{t-t_0}{2} A^{-1}d , \qquad y = y_0 + \frac{t-t_0}{2} A^{-1}d ,$$

i.e., our regime is a uniform rectilinear shift of the segment in direction perpendicular to the segment.

If $\varepsilon = -1$, then $d = 0$ and therefore $n_u = -n_v$ for any $t \in \Delta$. Equations (36) give us $\dot{n}_u = +pr$, $\dot{r} = -A^{-1}n_v$. We represent r as $r = e^{\phi j}$, where ϕ is Lipschitzian. Then from $|n_v|^2 \equiv 1$ it follows that

$$\dot{\phi} = \frac{\varepsilon}{|je^{\phi}|_A} ,$$

where $\varepsilon = \pm 1$. From $|r|^2 \equiv 1$ we get

$$\frac{d}{dt} |r|^2 = -2(n_v, r)_{A^{-1}} \equiv 0$$

and further

$$\frac{d}{dt} (n_v, r)_{A^{-1}} = p|r|^2_{A^{-1}} - |A^{-1}n_v|^2 \equiv 0 .$$

Whence

$$p = \frac{|A^{-1}n_v|^2}{|r|^2_{A^{-1}}} .$$

We obtain the following representation of this regime:

$$x = x_0 + \int_{t_0}^{t} \alpha A^{-1} n_u \, d\tau , \quad y = y_0 + \int_{t_0}^{t} \beta A^{-1} n_v \, d\tau ,$$

$$n_u = -n_v = - \frac{Aje^{\phi j}}{|je^{\phi j}|_A} ,$$

where α, β are measurable functions and $\alpha \geq 0$, $\beta \geq 0$, $\alpha + \beta = 1$.

If we parametrize this extremal using the sum of the lengths of the paths measured in the same metric as the length of the segment then it turns out that these are just precisely the motions under which the segment rotates in one direction, the instantaneous center of rotation being on it.

Now consider the sliding regime on $[a,b]$. As always, $p \equiv 0$ and $(r,\dot{x}) \neq 0$ in (a,b). Hence n_u, $n_v \equiv$ const. and $(r, A^{-1}n_u) \neq 0$ in (a,b). This means that in such a regime the endpoints of the segment slide along lines with directional vectors $A^{-1}n_u$, $A^{-1}n_v$, the sliding being bounded by the positions where the segment becomes orthogonal to one of the lines:

$$(r,\ A^{-1}n_u) = 0 \qquad \text{on} \quad (r,\ A^{-1}n_v) = 0 \ .$$

If $n_u = n_v$, then the regime is a uniform rectilinear shift in direction nonorthogonal to the segment.

We see that the parametrization is the only difference between the sliding regime in our problem and the sliding regime in the planar problem with a single metric.

Note that for a given d the durations of complete sliding can take on at most two values $(n = 2)$. This follows from the fact that n_u, n_v are determined uniquely to within a permutation.

Consider the rotating regime about the endpoint "x". The equations (30) take the form:

$$\dot{n}_u = -pr \ , \qquad \dot{n}_v = pr \ , \qquad \dot{y} = A^{-1}n_v \ , \tag{38}$$

with

$$|r|_{A^{-1}} \equiv 1 \ , \qquad |\dot{y}|_{A^{-1}} \equiv 1 \quad \text{and} \quad |n_u|_{A^{-1}} \leq 1 \ .$$

Consequently,

$$\frac{d}{dt}\,|r|^2 = -2(r,n_v)_{A^{-1}} \equiv 0$$

and

$$\frac{d}{dt}\,(n_v,r)_{A^{-1}} = p|r|^2_{A^{-1}} - |A^{-1}n_v|^2 \equiv 0 \ .$$

Whence

$$p = \frac{|A^{-1}n_v|^2}{|r|^2_{A^{-1}}} \ .$$

Thus n_v, r satisfy the closed system

$$\dot{n}_v = \frac{|A^{-1} n_v|^2}{|r|^2_{A^{-1}}} r , \qquad \dot{r} = -A^{-1} n_v ,$$

where

$$|n_v^0|_{A^{-1}} = |r_0| = 1 , \qquad (n_v^0, r_0)_{A^{-1}} = 0 .$$

It is easy to see that the subspace spanned by $n_v(t)$, $r(t)$, where t ranges over $[a,b]$, coincides with the minimal A-invariant subspace containing n_v^0, r_0. Hence for $n \geq 3$ one can construct nonplanar rotations of the segment.

For $n = 2$ it is easy to give a simple description of rotation. We write $r = e^{\phi j}$, where ϕ is Lipschitzian. Now from equations (38) just as in analyzing the degenerate regime with $d = 0$, we derive the representation of rotations:

$$x \equiv x_0 , \qquad y = x_0 - e^{\phi j} ,$$

$$n_u = d - n_v , \qquad n_v = \frac{Aje^{\phi j}}{|je^{\phi j}|_A} , \tag{39}$$

$$p = \frac{1}{|je^{\phi j}|^2_A \, |e^{\phi j}|^2_{A^{-1}}} ,$$

where ϕ satisfies the equation

$$\dot{\phi} = \frac{\varepsilon}{|je^{\phi j}|_A} , \qquad \varepsilon = \pm 1 \quad \text{and} \quad |n_u| \leq 1 .$$

Consider complete rotation on $[a,b]$. Then, since at a, b one has $|n_u|_{A^{-1}} = |n_v|_{A^{-1}} = 1$, then $n_u, n_v|_a = n_v, n_u|_b$ or $n_u, n_v|_a = n_u, n_v|_b$. Let us show that the latter is impossible. To this end, it suffices to see that $|\phi(b) - \phi(a)| < \pi$. If we had $|\phi(b) - \phi(a)| \geq \pi$ we could find $t^* \in [a,b]$, where $|\phi(t^*) - \phi(a)| = \pi$. Then according to the representation (39), $n_v(t^*) = -n_v(a)$. Hence

$$|n_u(t^*)| = |d - n_v(t^*)| = |d + n_v(a)| > 1 .$$

Contradiction. Thus,

$$n_u, n_v|_a = n_v, n_u|_b .$$

From the above arguments we obtain that any regime is either a uniform rectilinear shift of the segment in some direction, or motions of the segment in the plane, under which the segment rotates in one direction with the instantaneous centers of rotation on it; or a regime is a succession of combinant complete slidings and rotations about alternating endpoints.

Let us show that the last type of regime also after appropriate re-parametrization gives an extremal of the problem with one metric.

We have already seen that complete slidings possess such a property. Hence it suffices to see that complete slidings and rotations combine in our problem geometrically just as in the problem with one metric.

Suppose that there are combinant complete sliding and rotation on $[a,b]$, $[b,c]$. For definiteness, let $r_b \perp A^{-1}n_v(b)$. Then the rotation is about "x" and therefore

$$r_c \perp A^{-1}n_v(c), \qquad n_u, n_v|_b = n_v, n_u|_c$$

or

$$A^{-1}n_u(b) \perp r_c .$$

Since the sliding is along lines with directional vectors $A^{-1}n_u$, $A^{-1}n_v$, this means that the segment rotates by the same angle as in the problem with one metric, to position perpendicular to the line along which the fixed endpoint slid before this. Thus the property of interest to us is proved.

For $n \geq 3$, as we have seen, nonplanar rotations are possible, of which there are none in the problem with one metric. Consequently, the property of independence of extremals on the second metric occurs only on the plane.

• • • • •

CHAPTER 5

SYNTHESIS OF OPTIMAL MOTIONS OF SEGMENTS IN THE SPACE AND ON THE PLANE

The preceding part of the work was devoted to describing the collection of all extremals in a number of concrete particular versions of Ulam's problem. We will use this description to solve the problem of synthesizing the optimal motion for specified boundary positions of the segment. We strict ourselves to Ulam's problems in the space E^n and on the plane (no conditions on the endpoints).

The essence of the optimal synthesis problem is to remove from the list of extremals those which have no hope of optimality. The situation is complicated by the fact that in the given problem, due to the presence of the nonsmooth restriction $|u| + |v| \leq 1$, the available theory of Jacobi conjugate foci is inapplicable. So, what we need then is to find special ways of removing extremals which take direct account of the nature of Ulam's problem.

In the spatial case optimal synthesis leads to elementary

calculations. In the plane case we are successful in describing completely the extremals which yield an optimum. Construction of the optimal synthesis is reduced to an atlas which makes it possible to construct the desired optimal motion through ruler-and-compass constructions.

§ 1. GEOMETRIC PROPERTIES OF EXTREMALS

For purposes of studying geometric properties of extremals, let us write out differential equations for the support functions n_u, n_v. Equations (20) in our case assume the form

$$\dot{n}_u = -pr , \qquad \dot{n}_v = pr . \tag{40}$$

From (40) it follows at once that

$$n_u + n_v \equiv d = \text{const}.$$

Let us show that every extremal x, y assumes values in some three-dimensional affine manifold.

PROPOSITION 1. Let x, y be an extremals on $[t_0, t_1]$. There exists an affine manifold M, $\dim M \leq 3$, such that $x(t), y(t) \in M$ for all $t \in [t_0, t_1]$.

Proof. Fill out x, y to a stationary set x, y, n_u, n_v, p. Noting that n_u, n_v are supporting to the Euclidean norm $|\cdot|$ at $\dot{x}, \dot{y}$ we get

$$\dot{x} = |\dot{x}| n_u , \qquad \dot{y} = |\dot{y}| n_v ,$$

$$\dot{n}_u = -pr , \qquad \dot{n}_v = pr .$$

Whence

$$\dot{r} = |\dot{x}| n_u - |\dot{y}| n_v ,$$

i.e., the functions r, n_u, n_v satisfy a closed system of linear differential equations with measurable scalar coefficients. Hence for all $t \in [t_0, t_1]$ one has the inclusion

$$r(t),\ n_u(t),\ n_v(t) \in V ,$$

where V is the linear span of $r(t_0),\ n_u(t_0),\ n_v(t_0)$.

From the inclusion it follows that for any $t \in [t_0, t_1]$

$$x(t) = x(t_0) + \int_{t_0}^{t} |\dot{x}| n_u \, d\tau \in x(t_0) + V ,$$

$$y(t) = x(t_0) + \left(-r(t_0) + \int_{t_0}^{t} |\dot{y}| n_v \, d\tau \right) \in x(t_0) + V .$$

Thus the extremal x, y assumes values in the affine manifold $M = x(t_0) + V$, whose dimension is at most three.

This property of extremals makes it possible to limit ourselves to three-space in analyzing the spatial Ulam problem.

Now we establish an important property of extremals which consist of a series of alternating slidings along nonparallel lines and rotations about fixed endpoints. For brevity below, we will refer to such extremals as components.

PROPOSITION 2. Let x, y be a component extremal of Ulam's problem in E^n defined for all $t \in R$. There exists a motion D of E^n into itself, such that for any segment $[t_0, t_1]$ containing a sequentially running complete sliding and rotation, for any $t \in [t_0, t_1]$ one has

$$x(t + t_1 - t_0) = Dy(t) , \qquad y(t + t_1 - t_0) = Dx(t) .$$

For $n = 2$ this motion D is a shift. In other words, an extremal on a section including a complete sliding and rotation is obtained from the extremal on the preceding such section by means of composition of a permutation of the endpoints of the segment and the motion D.

Proof. Consider the stationary set x, y, n_u, n_v, p, including the component extremal x,y. Let us first of all show that the triple of vectors $n_u(t_1), n_n(t_1), r(t_1)$ can be obtained from the triple of vectors $n_v(t_0), n_u(t_0), -r(t_0)$ by means of an orthogonal transformation of E^n into itself. For this it suffices to show that the Grammians of these triples are equal.

Let t^* be a point in (t_0,t_1), such that on $[t_0,t^*]$ the regime is a complete sliding, and on $[t^*,t_1]$ a complete rotation. By formulas (23.4) and (26.4) one has

$$\kappa_{t_0} = \kappa_{t^*} , \qquad \chi_{t_0} = -\chi_{t^*} , \qquad \theta_{t_0} = \theta_{t^*}$$

and

$$\kappa_{t^*} = -\kappa_{t_1} , \qquad \chi_{t^*} = -\chi_{t_1} , \qquad \theta_{t^*} = \theta_{t_1} .$$

So, since

$$\kappa = \sigma + \gamma , \qquad \chi = \sigma - \gamma$$

we deduce that

$$\sigma_{t_1} = -\gamma_{t_0} , \qquad \gamma_{t_1} = -\sigma_{t_0} , \qquad \theta_{t_1} = \theta_{t_0} .$$

In other words,

$$(n_u(t_1), r(t_1)) = (n_v(t_0), -r(t_0)) ,$$

$$(n_v(t_1), r(t_1)) = (n_u(t_0), -r(t_0)) ,$$

$$(n_u(t_1),\ n_v(t_1)) = (n_v(t_0),\ n_u(t_0)) .$$

Since the vectors $n_u(t_0)$, $n_v(t_0)$, $r(t_0)$, $n_u(t_1)$, $n_v(t_1)$, $r(t_1)$ all have unit length, this proves the Grammians coincide. Hence, there is an orthogonal transformation $A: E^n \to E^n$ such that

$$n_u(t_1) = An_v(t_0) , \quad n_v(t_1) = An_u(t_0) , \quad r(t_1) = -Ar(t_0) .$$

It is now easy to check that the functions

$$\bar{x}(t) = c + Ay(t-\Delta) , \quad \bar{n}_u(t) = An_v(t-\Delta) ,$$

$$\bar{y}(t) = c + Ax(t-\Delta) , \quad \bar{n}_v(t) = An_u(t-\Delta) ,$$

$$\bar{p}(t) = p(t-\Delta) ,$$

where

$$\Delta = t_1 - t_0 , \quad c = x(t_1) - Ay(t_0)$$

form a stationary set extending the set x, y, n_u, n_v, p from the segment $[t_0, t_1]$ to the segment $[t_1, t_1+\Delta]$. Since the extension is unique, both sets coincide for $t \in [t_1, t_1+\Delta]$. Thus,

$$x(t) = Dy(t-\Delta) , \quad y(t) = Dx(t-\Delta) ,$$

where $Dz = c + Az$ is the desired motion of E^n into itself. For $n = 2$,

$$n_u(t_1) = n_v(t_0) , \quad n_v(t_1) = n_u(t_0)$$

and consequently $A = E$ and D is a shift. For $n \geq 3$ the orthogonal transformation can be chosen so that its determinant $\det A = 1$.

From Proposition 2 it follows that the angles of planar rotations of the segment about fixed endpoints in a component extremal are all equal. As follows from (25), one can compute these angles by the formula

$$\Delta\phi = 2 \arccos \frac{|d|^2}{\sqrt{|d|^4 + 4(d, r(\tilde{t}))^2}}, \qquad (41)$$

where $r(\tilde{t})$ is computed for any time $\tilde{t}$ lying on the boundary of a complete sliding ($|(d, r(\tilde{t}))|$ has one value at all points $\tilde{t}$ of the boundary of a complete sliding on a component extremal).

For $n = 2$ formula (41) can be simplified:

$$\Delta\phi = \arccos \left(\frac{|d|^2}{2} - 1 \right) = \arccos \, (n_u(\tilde{t}), \, n_v(\tilde{t})) \, .$$

In other words, $\Delta\phi$ coincides with the angle between the lines along which the sliding takes place.

In concluding this point, let us give a complete description of extremals in the spatial problem. From Theorem 2, noting what has been proved, it follows that the extremals make up the following list:

- 1. parallel displacement of the segment;
- 2. component motion consisting of sequentially running slidings of the segment along nonparallel lines and planar rotations about fixed alternating endpoints by a constant angle $\Delta\phi$. Here the arcs of the rotations are tangent to the corresponding sliding lines, and the functions $\dot{x}(t)$, $\dot{y}(t)$ are continuous functions of the parameter t (the sum of the lengths of the

paths swept out by the endpoints of the segment).

The angle $\Delta\phi$ is computed by (41);

- 3. planar motion in which the angular rate of rotation of the segment is constant, and the instantaneous center of rotation of the segment, as a solid body, lies on it all the time;
- 4. helical motion of the segment in which the midpoint of the segment moves uniformly and rectilinearly, and the segment itself rotates uniformly in a plane orthogonal to the direction of motion of the midpoint.

(42)

We will call an extremal of type (3.) a planar singular motion.

In view of this list of extremals it follows, in particular, that every extremal containing a section of parallel displacement is an extremal of type (1.), i.e., a parallel displacement. Meanwhile, in some works devoted to Ulam's problem (see [4]) there are erroneously described extremals on which parallel displacement is replaced by rotation of the segment about a fixed endpoint.

§ 2. OPTIMAL MOTION OF SEGMENTS IN THE SPACE E^3

We begin the construction of optimal synthesis in E^3 with a description of the mutual arrangement of a pair of segments that allows for the most economical motion -- a sliding of the endpoints over a pair of lines. We call this mutual arrangement articulated.

Let us set down some notation. Let x_0, y_0; x_1, y_1 be the boundary positions of the segments; let e_1, e_2 be the unit directional vectors of the lines along which the endpoints x, y are to slide, and let

$$\alpha_1 = (x_1 - x_0,\ e_1) > 0 , \qquad \beta_1 (y_1 - y_0,\ e_2) > 0 .$$

Also we write $\theta = (e_1, e_2)$.

We have to establish a criterion for the existence of a pair of nondecreasing continuous functions $\alpha(t)$, $\beta(t)$, $t \in [0,1]$, such that

- a. $\alpha(0) = 0$; $\beta(0) = 0$, $\alpha(1) = \alpha_1$, $\beta(1) = \beta_1$;
- b. $|x(t) - y(t)| \equiv 1$,

(43)

where

$$x(t) = x_0 + \alpha(t) e_1 , \qquad y(t) = y_0 + \beta(t) e_2 .$$

<u>PROPOSITION 3</u>. The arrangement of the segments x_0, y_0; x_1, y_1 is articulated if and only if the numbers $(x_1 - x_0,\ r_0)$, $(x_1 - x_0,\ r_1)$, $(y_1 - y_0,\ r_0)$, $(y_1 - y_0,\ r_1)$ have the same sign.

<u>Proof</u>. Necessity. Thus, assume that the arrangement of x_0, y_0; x_1, y_1 is articulated, i.e., there are functions $\alpha(t)$, $\beta(t)$, satisfying conditions (43). We have to show that all the pairwise products (e_i, r_j), $i = 1,2$, $j = 0,1$, are nonnegative. We argue by contradiction.

Suppose not. Then one of the following occurs.

- 1. $(e_1, r_0)(e_2, r_0) < 0$.

  Consider here the identity $|r(t)|^2 \equiv 1$. We have

$$|r(t)|^2 - 1 = \beta^2 + \alpha^2 + 2\beta(e_2, r_0) - 2\alpha(e_1, r_0) - 2\alpha\beta(e_1, e_2) .$$

Let

$$t^* = \inf \{t>0 \mid \alpha(t) + \beta(t) > 0\} .$$

Then for sufficiently small $\delta > 0$ on the interval $(t^*, t^*+\delta)$ the number $|r(t)|^2 - 1$ will be either strictly negative or strictly positive, which contradicts $|r(t)|^2 \equiv 1$.

• 2. $(e_1,r_0)(e_1,r_1) < 0$.

Note that in this case $|\theta| < 1$, for otherwise $e_1 = e_2$, $r(t) \equiv$ const., whence $(e_1,r_0)(e_1,r_1) = (e_1,r_0)^2 \geq 0$. Further, since the continuous function $(e_1, r(t))$ takes values of alternate signs for $t = 0,1$, we can find t, $0 < t < 1$, such that $(e_1, r(t)) = 0$. For this t one has

$$\begin{aligned} 0 &= (e_1, r(t))^2 \\ &= (\theta^2 - 1)\beta^2(t) + 2\beta(t)[(e_2,r_0) - \theta(e_1,r_0)] + (e_1,r_0)^2 . \end{aligned} \tag{44}$$

On the other hand, note that the function

$$\Delta(\beta) = (\theta^2 - 1)\beta^2 + 2\beta[(e_2,r_0) - \theta(e_1,r_0)] + (e_1,r_0)^2$$

is strictly convex in β and $\Delta(0) \geq 0$, $\Delta(\beta_1) \geq 0$. Hence, $\Delta(\beta(t)) > 0$, which contradicts (44).

The remaining cases

$$(e_2,r_0)(e_2,r_1) < 0 , \qquad (e_1,r_1)(e_2,r_1) < 0$$

are examined similarly. The necessity is proved.

Sufficiency. Now assume that the (e_i,r_j) have the same sign. We take α as a parameter with values on $[0,\alpha_1]$, and we will seek a function $\beta(\alpha)$ such that

$$|x_0 + \alpha e_1 - y_0 - \beta(\alpha)e_2|^2 \equiv 1 ,$$

i.e.,

$$\beta^2(\alpha) + 2\beta(\alpha)(e_2, r_0-\alpha e_1) + |r_0 - \alpha e_1|^2 - 1 = 0 .$$

Simple analysis reveals that the formula

$$\beta(\alpha) = (e_2, \alpha e_1-r_0) + \operatorname{sign}(e_2,r_0)\sqrt{(e_2, r_0-\alpha e_1)^2 - |r_0 - \alpha e_1|^2 + 1}$$

determines the desired nondecreasing function $\beta(\alpha)$, and $\beta(0) = 0$, $\beta(\alpha_1) = \beta_1$. Q.E.D.

The above articulated arrangement criterion allows one to characterize the optimal component extremals.

PROPOSITION 4. Let x, y be a component extremal which on $[t_0,t_1]$ contains two complete rotations and a sliding between them. Then the segments $x(t_0), y(t_0)$; $x(t_1), y(t_1)$ are arranged in articulated fashion.

Proof. The proof is easy, using Proposition 3. From Proposition 4 it follows that an optimal, that is a component extremal, cannot include two complete rotations and may include at most two complete rotations. In fact, otherwise we can replace it on a minimal segment containing a pair of complete rotations by a sliding over a pair of lines. As a result of the change we get a way of moving segments that is more economical than the original component extremal -- a contradiction to its optimality.

We will call the arrangement of a pair of segments in E^3 common if they are non-coplanar and non-orthogonal simultaneously to the line through their midpoints. In this case the optimal motion is a priori an extremal of component type since the other types of extremals in the list (42) cannot satisfy the boundary conditions. Proposition 4 reduces the construction of an optimal to calculating the parameters of a component extremal including at most two complete rotations and satisfying the boundary conditions. We are not going to write the corresponding systems of equations as they involve a routine but very cumbersome analysis.

If the arrangement of a pair of segments is not common, then in constructing an optimal, besides the component motion, it is necessary to consider extremals of types (1.), (3.), (4.) from the list (42) and for each type to calculate the best extremal moving the segments. The construction of an optimal singular motion is described in detail in Section 3 below. It is proved that if $\max \{|x_0-y_1|, |y_0-x_1|\} > 1$, then a construction of an optimal singular motion makes no sense, for there exists a more economical planar component extremal. An optimal helical motion is characterized by the fact that the angle of rotation of the segment in the plane orthogonal to the axis of helical motion is equal to $\arccos (r_0, r_1)$, i.e., to the angle between the endpoint positions of the oriented segment.

§ 3. ATLAS OF OPTIMAL MOTIONS OF SEGMENTS ON THE PLANE

In the case of the plane E^2, optimal motion of segments can be synthesized by simple geometric constructions. We will compose a scheme permitting these constructions for an arbitrary pair of segments. Such schemes are called atlases.

Note that in the case of the plane the list of extremals (42) is abridged and includes only types (1.), (2.), (3.). Let us establish some simple geometric properties of planar extremals. Fix on E^2 an orthonormal basis i, j. For every pair of continuous functions x, y describing the movement of the endpoints of a unit segment in the plane we will let $\phi(t)$ denote a continuous function such that

$$r(t) = e^{\phi(t)j} .$$

PROPOSITION 5. Let x, y be a planar extremal. Then ϕ is a continuously differentiable function, and $\dot{\phi}(t)$ does not change sign.

Proof. Thus, let x, y be some planar extremal. If the extremal is a parallel displacement or a singular motion, then there is nothing to prove since ϕ is a linear function. It remains only to consider the case where our extremal is component. Noting that

$$\dot{r} = j\dot{\phi}e^{\phi j} \quad \text{and} \quad \dot{r} = \dot{x} - \dot{y} = |\dot{x}|n_u - |\dot{y}|n_v ,$$

we get the representation

$$\dot{\phi} = -je^{-\phi j}(|\dot{x}|n_u - |\dot{y}|n_v) .$$

Since on a component extremal the derivatives $\dot{x}, \dot{y}$ are continuous (because the stationary set including the extremal does not degenerate at any instant), $\dot{\phi}$ is continuous. Now let us show that $\dot{\phi}(t)$ does not vanish anywhere, i.e., $|\dot{x}|n_u - |\dot{y}|n_v \neq 0$. Indeed, if $|\dot{x}(t)||\dot{y}(t)| > 0$, then $|\dot{x}|n_u - |\dot{y}|n_v \neq 0$ since $n_u(t), n_v(t)$ are linearly independent vectors. But if

$|\dot{x}(t)||\dot{y}(t)| = 0$, then either $\dot{x}(t) = 0$ and therefore $|\dot{y}(t)| = 1$, $|n_v(t)| = 1$, or $\dot{y}(t) = 0$ and therefore $|\dot{x}(t)| = 1$, $|n_u(t)| = 1$. Thus, for any t one has $\dot{\phi}(t) \neq 0$, so by continuity $\dot{\phi}(t)$ does not change signs. Proposition 5 is proved.

Thus, in moving along a plane extremal the segment rotates strictly in one direction. Let us consider in more detail the singular plane extremals. These extremals can be represented in the form:

$$x(t) = x_0 + \int_0^t \varepsilon j\, \alpha(\tau) e^{(\varepsilon\tau+\phi_0)j}\, d\tau ,$$

$$y(t) = y_0 - \int_0^t \varepsilon j\, \beta(\tau) e^{(\varepsilon\tau+\phi_0)j}\, d\tau ,$$

where ε is an integral parameter $(\varepsilon = \pm 1)$; $\alpha(t)$, $\beta(t)$ are nonnegative measurable functions and $\alpha + \beta = 1$.

For every singular extremal $\dot{\phi}(t) \equiv \varepsilon$, whence

$$t_1 = |\phi(t_1) - \phi(0)| = \int_0^{t_1} (|\dot{x}| + |\dot{y}|)\, d\tau .$$

Since for arbitrary Lipschitzian motion of the unit segment the inequality

$$\int_0^{t_1} (|\dot{x}| + |\dot{y}|)\, d\tau \geq \int_0^{t_1} |\dot{\phi}|\, d\tau \geq |\phi(t_1) - \phi(0)|$$

is satisfied, a singular extremal, for which $|\phi(t_1) - \phi(0)| \leq \pi$ is an optimal motion. Hence it is worth interest to describe the arrangements of segments for which there exists a moving singular extremal with complete rotation not exceeding π.

<u>PROPOSITION 6</u>. The segment x_0, y_0 admits motion with the segment x_1, y_1 by means of a singular extremal with complete rotation $\leq \pi$ if and only if $|x_0 - y_1| \leq 1$, $|y_0 - x_1| \leq 1$. A singular extremal moving the given segments can be in the form of no more than three rotations about fixed endpoints of the segment.

<u>Proof</u>. Necessity. Thus, suppose there is a singular extremal $x(t)$, $y(t)$, $t \in [0,t_1]$ such that $x(0) = x_0$, $y(0) = y_0$, $x(t_1) = x_1$, $y(t_1) = y_1$, $|\phi(t_1) - \phi(0)| \leq \pi$. Since for a singular extremal $\phi(t) = \varepsilon t + \phi_0$, the last condition gives $t_1 \leq \pi$.

Taking a suitable orthonormal basis i, j in E^2, we get the following representation of a singular extremal

$$\begin{aligned} x(t) &= x_0 + \int_0^t \alpha(\tau)\, j\, e^{\tau j}\, d\tau , \\ y(t) &= y_0 - \int_0^t \beta(\tau)\, j\, e^{\tau j}\, d\tau , \end{aligned} \tag{44}$$

with

$$x_0 - y_0 = i .$$

Our goal is to prove $|x(t_1) - y_0| \leq 1$, $|y(t_1) - x_0| \leq 1$. Using (44), these inequalities become

$$\begin{aligned} \left| i + \int_0^{t_1} \alpha(\tau)\, j\, e^{\tau j}\, d\tau \right| &\leq 1 , \\ \left| i + \int_0^{t_1} \beta(\tau)\, j\, e^{\tau j}\, d\tau \right| &\leq 1 . \end{aligned} \tag{45}$$

To prove (45) it suffices to show that for any $\mu \in [-\pi,\pi]$ and any measurable function $0 \le u(t) \le 1$ we have the inequality

$$\left(e^{\mu j},\quad i + \int_0^{\pi} u(\tau)\, j\, e^{\tau j}\, d\tau\right) \ge -1$$

or, equivalently,

$$J(u) = \cos \mu + \int_0^{\pi} u(\tau) \sin(\mu - \tau)\, d\tau \ge -1 .$$

To prove the last inequality, we consider the auxiliary optimal control problem

$$J(u) \to \min , \qquad 0 \le u \le 1 .$$

Applying Pontryagin's maximum principle, it is easy to show that

$$\min J(u) = \begin{cases} -1 & \text{for } \mu \in [0, \pi] \\ -\cos \mu & \text{for } \mu \in [-\pi, 0] \end{cases} .$$

Thus for any μ, u one has $J(u) \ge -1$, which completes the proof of the necessity.

Sufficiency. Suppose that the boundary positions of the segments satisfy $|x_0 - y_1| \le 1$, $|y_0 - x_1| \le 1$. Fix an orientation on E^2 such that $0 \le (\widehat{r_0, r_1}) \le \pi$. One can show that there is a singular extremal consisting of three successive rotations in the positive direction about fixed endpoints: about x, then about y and then again about x, and $|y(t') - x_1| = 1$, $x(t'') = x_1$, $y(t_1) = y_1$. Here $0 \le t' \le t'' \le t_1 = (\widehat{r_0, r_1})$ are the instants separating the sections of rotations about differently labeled endpoints.

Indeed, on the first stage we will rotate the segment from initial position x_0, y_0 in the positive direction about x

until we get $|y - x_1| = 1$. The desired angle of rotation t' is determined as the root solution of the equation

$$|y(t') - x_1| = 1 ,$$

where

$$y(t) = x_0 + e^{tj}(y_0 - x_0) .$$

The equation has a solution $0 \le t' \le \phi_1 - \phi_0$ since $y(t)$ is continuous and the inequalities $|y(0) - x_1| \le 1$, $|y(\phi_1-\phi_0) - x_1| \ge 1$ are satisfied. The last inequality follows from the well-known geometrical "parallelogram" identity:

$$|y(\phi_1-\phi_0) - x_1|^2 + |x_0-y_1|^2 = 2 + |x_0-x_1|^2 + |y(\phi_1-\phi_0) - y_1|^2 ,$$

whence, since $|x_0-y_1| \le 1$, we get $|y(\phi_1-\phi_0) - x_1| \ge 1$.

Having determined t', we will rotate the segment in the positive direction about y until we have $x = x_1$. The instant t'' is $t'' = t' + \Delta\phi$, where $\Delta\phi$ is the desired angle of rotation. Finally, the third and final stage consists in rotating in the positive direction about x until coincidence with the final segment. On the thusly constructed singular extremal the complete rotation does not exceed π.

From Proposition 6 it follows that even though arbitrary measurable weight functions figure in the representation of singular extremals, any such extremal is equivalent (in the sense of the final positions of the segment and the sum of the lengths of the paths swept out by the endpoints) to a number of rotations about fixed endpoints.

THEOREM 4. In order that the plane extremal x, y moving

the segments $x_0, y_0; x_1, y_1$ be optimal it is necessary and sufficient that the complete rotation of the segment on it not exceed π.

The proof of this theorem is rather lengthy and will be omitted. Let us turn to a detailed description of synthesis of planar extremals moving given positions of segments. The optimality of the extremals constructed below follows from Theorem 4, because on all these extremals the complete rotation of the segment does not exceed π.

Thus, let there be given the initial position A, B and the final position A', B' of a segment. Let R and R' be the point sets of the segments, respectively. We begin a sequential analysis of all possible versions of the mutual positioning of two segments on the plane.

First consider the case $R \cap R' \neq \emptyset$. The following versions are possible.

• 1. Two endpoints with the same letter coincide. Here the required extremal is a singular motion -- this is a rotation about the common endpoint by an angle $\leq \pi$.

• 2. There are no endpoints with the same letter that coincide (Figure 12).

Here the following cases are possible:

• 2.1. $\max\{\angle A, \angle A', \angle B, \angle B'\} \geq \frac{\pi}{2}$.

• 2.2. $\max\{\angle A, \angle A', \angle B, \angle B'\} < \frac{\pi}{2}$.

Let us consider these cases in more detail.

• 2.1. Let $\angle A' \geq \frac{\pi}{2}$ (Figure 12). Let us additionally single out some other variants.

2.1.1. $\angle B \geq \frac{\pi}{2}$. Here the positioning of the segments is articulated and the desired extremal is a sliding of the endpoints of the segment along the lines AA', BB'.

2.1.2. $\angle B < \frac{\pi}{2}$. Here the extremal is a rotation of the segment about A to position AC and a subsequent sliding along the lines (Figure 15). The construction is feasible since from the fact that $\angle A' \geq \frac{\pi}{2}$ it follows that

$$\rho(A,B') > \rho(A',B') = 1 .$$

• 2.2. Here we isolate the following variants:

2.2.1. $\max \{\rho(A',B), \rho(A,B')\} > 1$.

2.2.2. $\max \{\rho(A',B), \rho(A,B')\} \leq 1$.

Let us consider these cases in more detail.

2.2.1. Let $\rho(A,B') > 1$. Then the extremal is a rotation of the segment AB about A to position AC, then a sliding to position C'B' and a rotation about B' from position C'B' to A'B'. The feasibility of the construction follows at once from the inequality $\rho(A,B') > 1$ (Figure 16).

2.2.2. According to Proposition 6, the optimum is realized on a singular extremal for which the total rotation of the segment is at most π. This singular extremal can be represented in the form of no more than three rotations about fixed endpoints. This construction is described in the proof of Proposition 6.

Thus, we have learned to solve the synthesis problem for intersecting segments. Let us pass to the case $R \cap R' = \emptyset$.

Here we single out the two possibilities:

• 1. R, R' lie on one side of the line AA' or BB' (a closed half-plane is intended) (Figure 14).

• 2. R, R' lie on different sides of the lines AA', BB' (Figure 13).

Let us examine these variants.

• 1. In this case the contour $ABB'A'$ defines a quadrangle (perhaps a degenerate one, i.e., lying on a line) (Figure 14). Take a pair of same-letter vertices of the quadrangle with sum of the interior angles $\leq \pi$. For definiteness, say A, A'. We additionally consider the following variants:

• 1.1. $\max \{\angle A, \angle A'\} \geq \frac{\pi}{2}$,

• 1.2. $\max \{\angle A, \angle A'\} < \frac{\pi}{2}$.

• 1.1. Here the positioning of the segments is specified and the desired extremal is a sliding along AA', BB'.

• 1.2. Here we need to also look at a couple of subcases:

1.2.1. $\min \{\angle B, \angle B'\} > \frac{\pi}{2}$. In this case the extremal is a sliding of AB into position A_1C, a rotation about C to position A_2C and then again a sliding to position $A'B'$ (Figure 17). The construction is feasible because $\angle B, \angle B' > \frac{\pi}{2}$ implies that $\rho(B,A'), \rho(A,B') > 1$.

1.2.2. $\min \{\angle B, \angle B'\} \leq \frac{\pi}{2}$. For definiteness, let

$\angle B' \le \frac{\pi}{2}$. Then the extremal is a sliding of AB into the position A_1B' and then a rotation about B' to position $A'B'$ (Figure 18).

• 2. The following situations are possible (Figure 13):

• 2.1. $\max\{\angle A, \angle B, \angle A', \angle B'\} \ge \frac{\pi}{2}$.

• 2.2. $\max\{\angle A, \angle B, \angle A', \angle B'\} < \frac{\pi}{2}$.

• 2.1. Let $\angle B \ge \frac{\pi}{2}$. We single out two further cases.

2.1.1. $\angle B' \ge \frac{\pi}{2}$. Here there are two extremals of component type, moving the segments AB, $A'B'$. One of them consists of a sliding of AB into A_1B_1, then follows a rotation about B_1 to position A_2B_1 and then a sliding to position $A'B'$ (Figure 19). The construction of this extremal is feasible since $\rho(A,B') > 1$, $\rho(B,A') > 1$. The other extremal is constructed similarly and consists of a sliding of AB to position A_1B_1, then a rotation about A_1 to position A_1B_2, and then a sliding to position $A'B'$.

These extremals differ in the direction of rotation of the segment. On the first extremal the total rotation of the segment is $\phi_1 = \pi + \angle B - \angle B'$ and on the second is $\phi_2 = \pi + \angle B' - \angle B$. Consequently, if $\angle B < \angle B'$ then $\phi_1 < \pi$, $\phi_2 > \pi$ and, by Theorem 4, the first extremal is optimal but the second is not. But if $\angle B \ge \angle B'$ then $\phi_2 \le \pi$, hence the extremal is optimal. Finally, if $\angle B = \angle B'$ then both extremals are equivalent and bring about optimal motion of the segments.

2.1.2. $\angle B < \frac{\pi}{2}$ (Figure 20). Here, if $\angle A_1B'B < \angle A'B'B$ then the extremal consists of a sliding of AB to position A_1B', then a rotation about B' to position A'B'. The construction of the extremal is feasible since $\rho(A,B') > 1$.

But if $\angle A_1B'B \geq \angle A'B'B$ then the arrangement of the segments is articulated and the desired extremal is a sliding along the lines AA', BB'.

• 2.2. We examine here additionally the cases:

2.2.1. $\max \{\rho(A,B'), \rho(B,A')\} > 1$.

2.2.2. $\max \{\rho(A,B'), \rho(B,A')\} \leq 1$.

2.2.1. For definiteness, let $\rho(A,B') > 1$. The segments AB, A'B' in this arrangement, generally speaking, can be moved by two component extremals differing in direction of rotation. The first extremal consists of a rotation of AB about A to position AB_1, then a sliding to position A_1B' and then a rotation about B' to position A'B' (Figure 21). The feasibility of the construction follows from the condition $\rho(A,B') > 1$.

If $\angle B \leq \angle B'$, then $\rho(A'B) \geq \rho(A,B') > 1$ and there is an analogous extremal consisting of a rotation of AB about B to position A_1B, then a sliding to position B_1A' and a rotation to A'B'.

In the first case the rotation of the segment is $\phi_1 = \pi + \angle B' - \angle B$, and in the second one, $\phi_2 = \pi + \angle B - \angle B'$. Consequently, if $\angle B \geq \angle B'$ then $\phi_1 \leq \pi$ and the optimal is the first extremal; but if $\angle B < \angle B'$ then $\phi_2 \leq \pi$ and the optimal is the second extremal. Finally, if $\angle B = \angle B'$ then both extremals are equivalent and effect optimal motion of the segments.

2.2.2. In this case, according to Proposition 6, optimal motion can be effected on a singular extremal with total rotation $\leq \pi$ consisting of no more than three rotations about fixed endpoints.

We have thus analyzed all possible cases of positioning two segments on the plane. In the figures we have clearly indicated the geometric ruler-and-compass constructions needed to get the optimal synthesis. The basic operation in these constructions is drawing the tangent from a given point to a given circle. For example, in Figure 15, the optimal motion consists of a rotation about A to position AC and then a sliding along the lines AA', CB' to position A'B'. The line CB' is constructed by drawing the tangent from B' to the unit circle centered at A, the line AA' passes directly through the points A, A'.

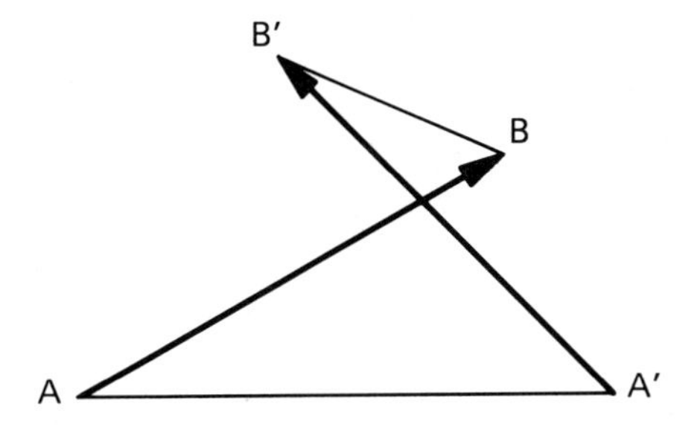

Figure 12

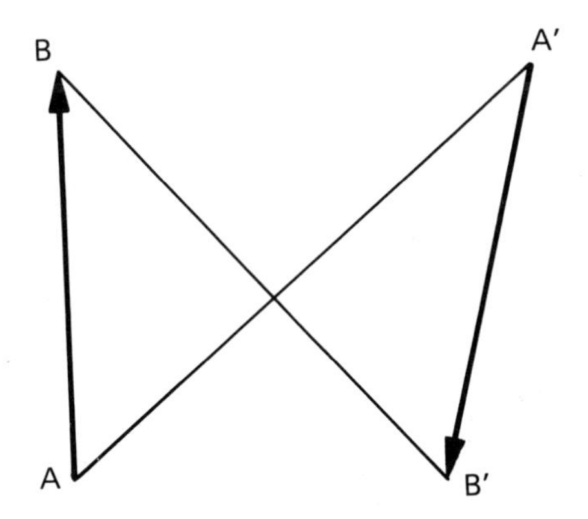

Figure 13

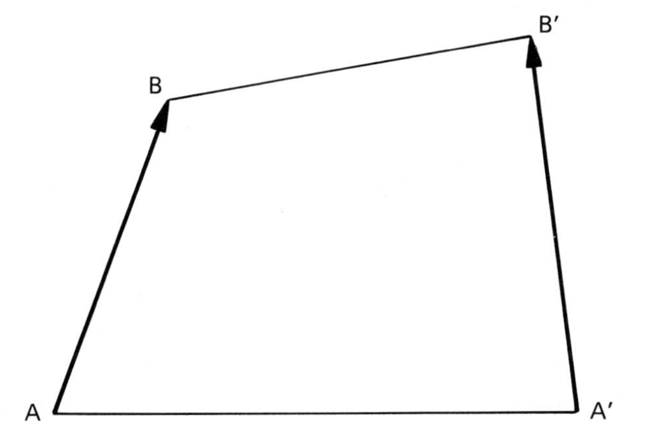

Figure 14

As we have seen, synthesis of optimal motion of segments requires a rather cumbersome and "branched-out" analysis. The analysis can be made using the block-diagram (below).

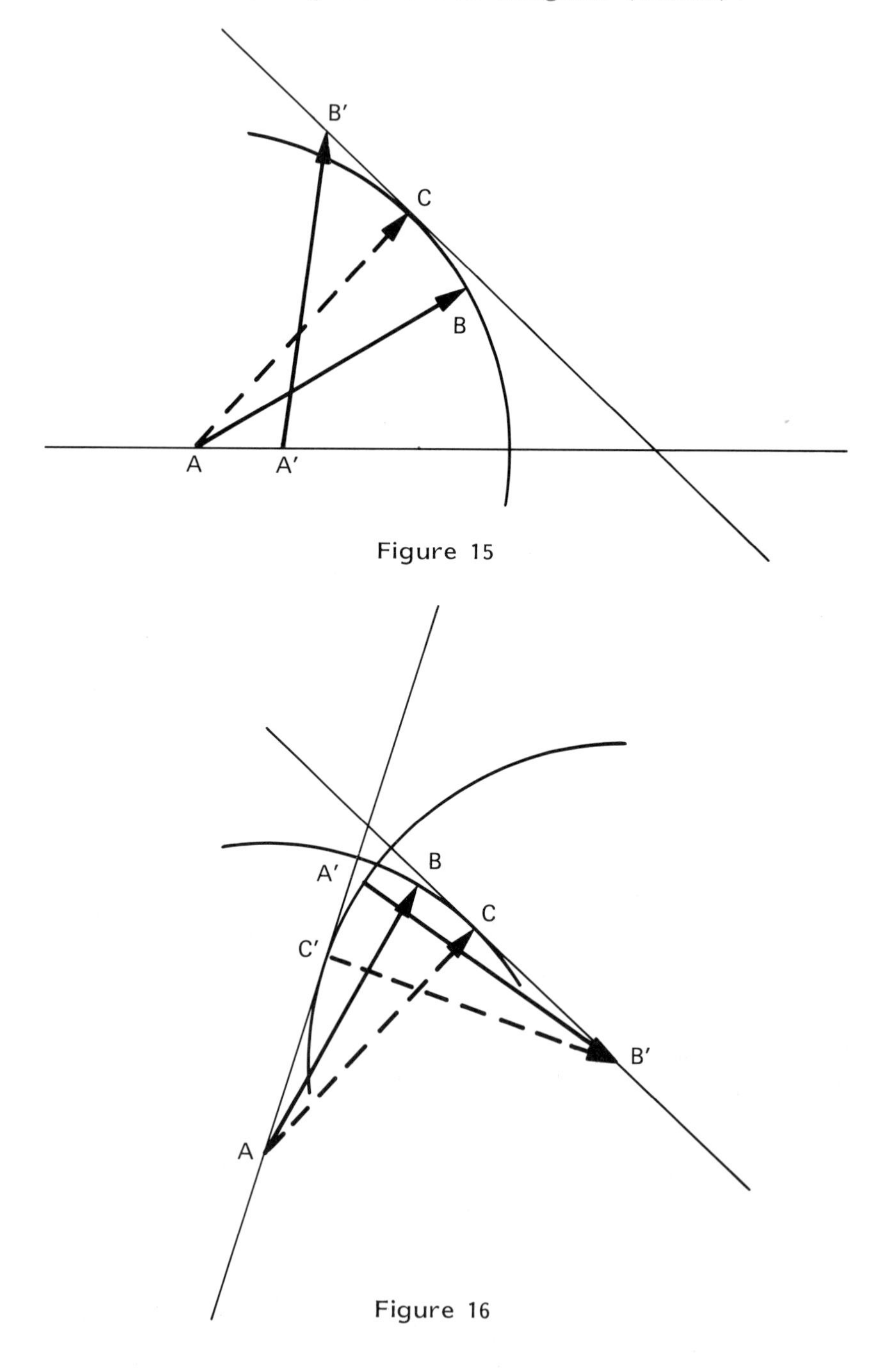

Figure 15

Figure 16

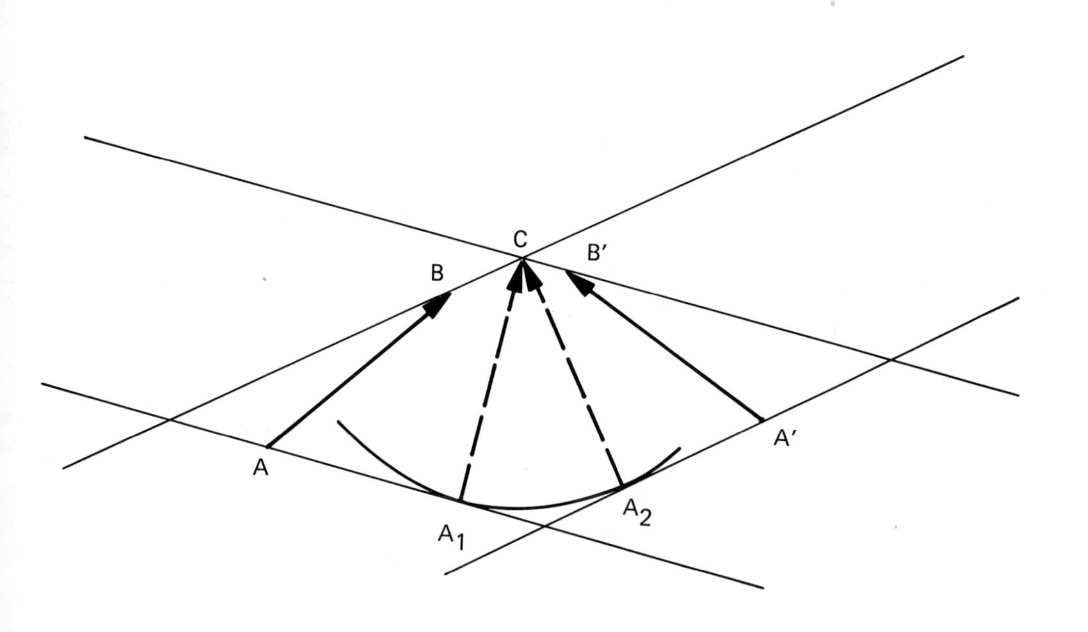

Figure 17

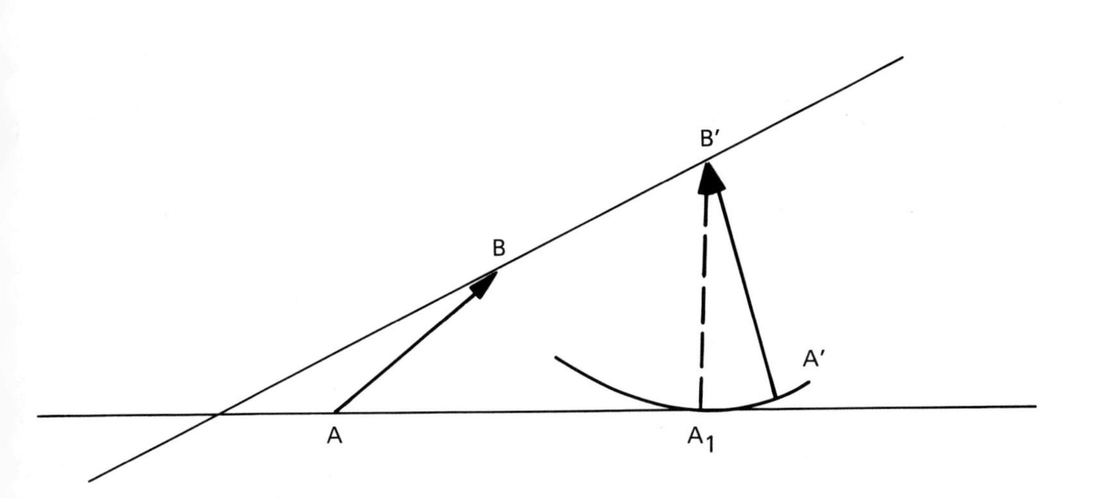

Figure 18

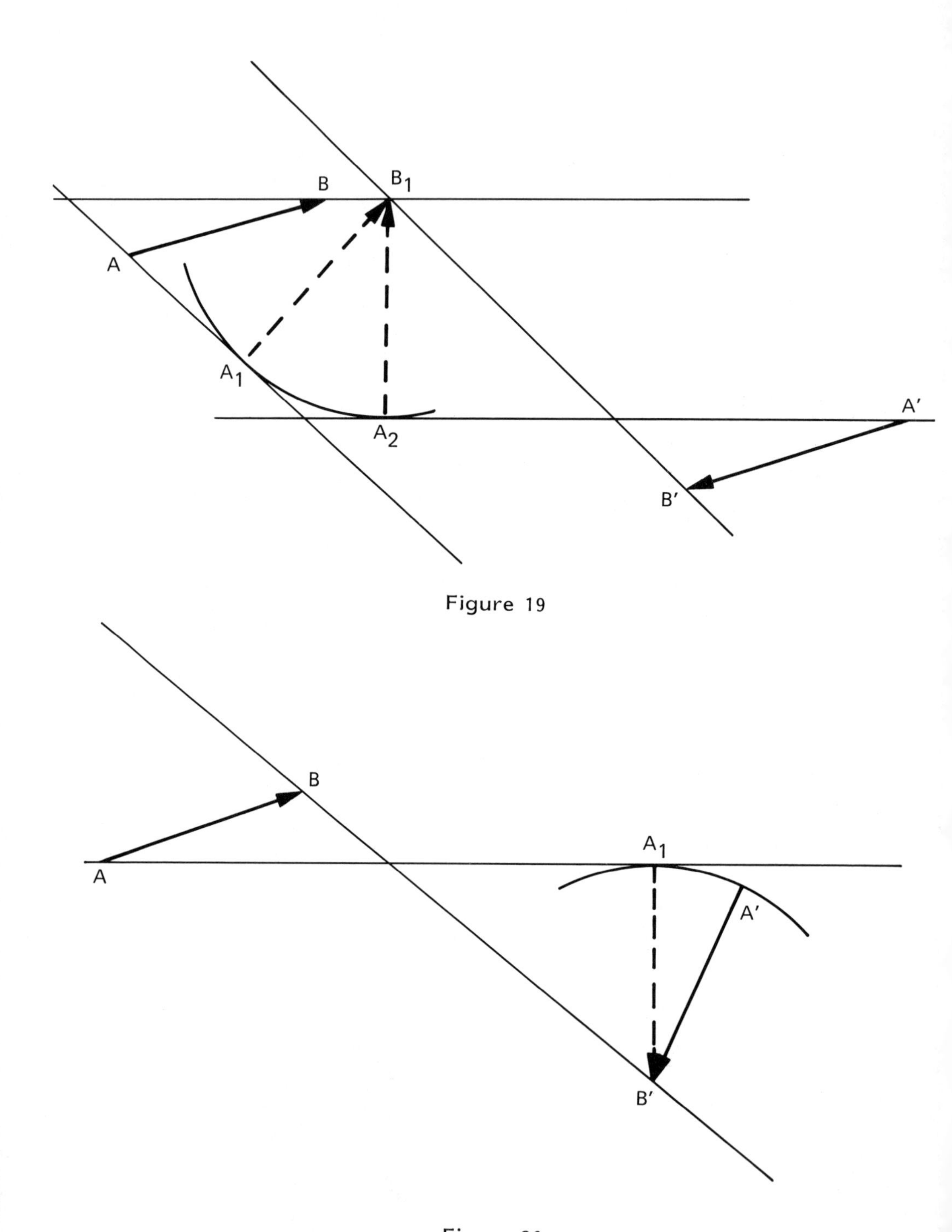

Figure 19

Figure 20

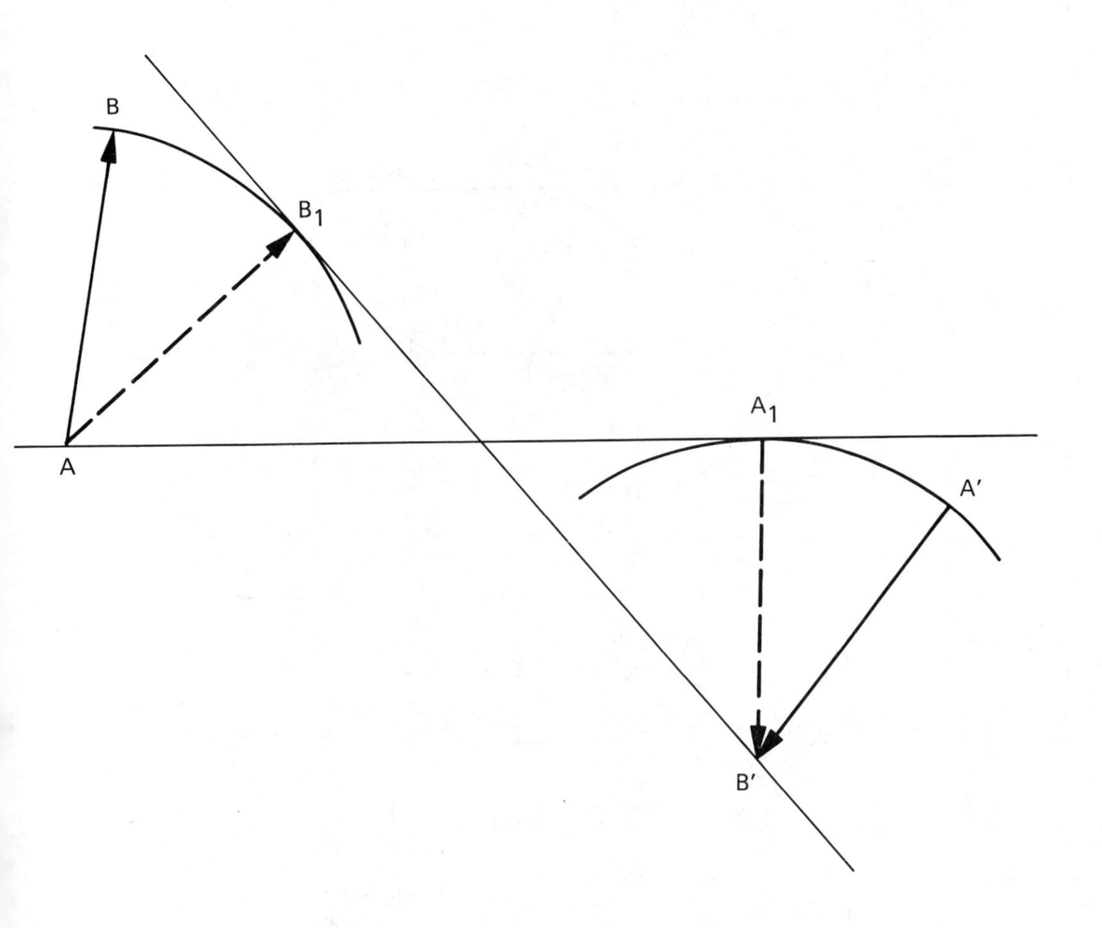

Figure 21

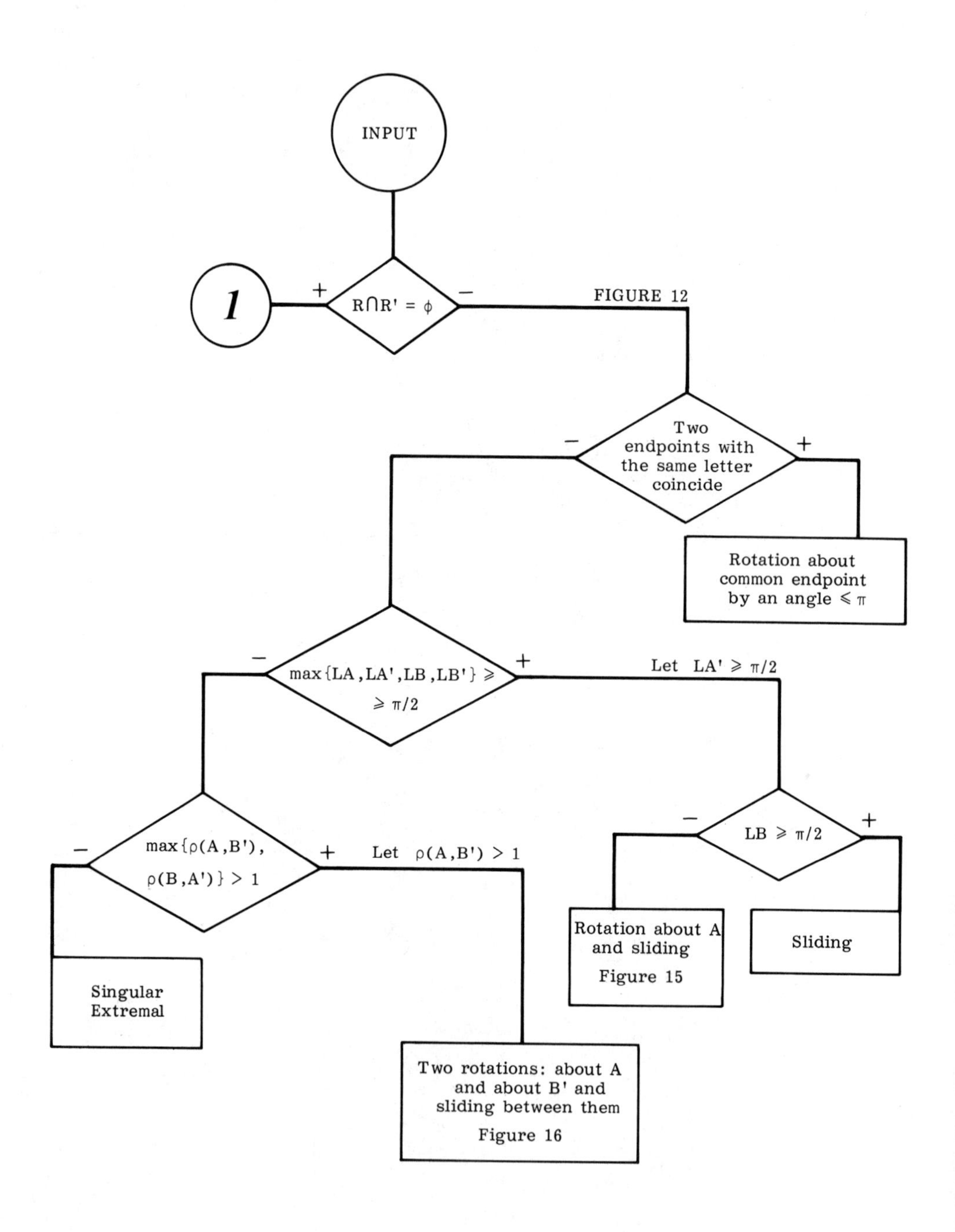
INPUT
1
+
R⋂R' = ϕ
−
FIGURE 12
Two endpoints with the same letter coincide
−
+
Rotation about common endpoint by an angle ⩽ π
max{LA,LA',LB,LB'} ⩾ ⩾ π/2
−
+
Let LA' ⩾ π/2
max{ρ(A,B'), ρ(B,A')} > 1
−
+
Let ρ(A,B') > 1
LB ⩾ π/2
−
+
Singular Extremal
Two rotations: about A and about B' and sliding between them
Figure 16
Rotation about A and sliding
Figure 15
Sliding

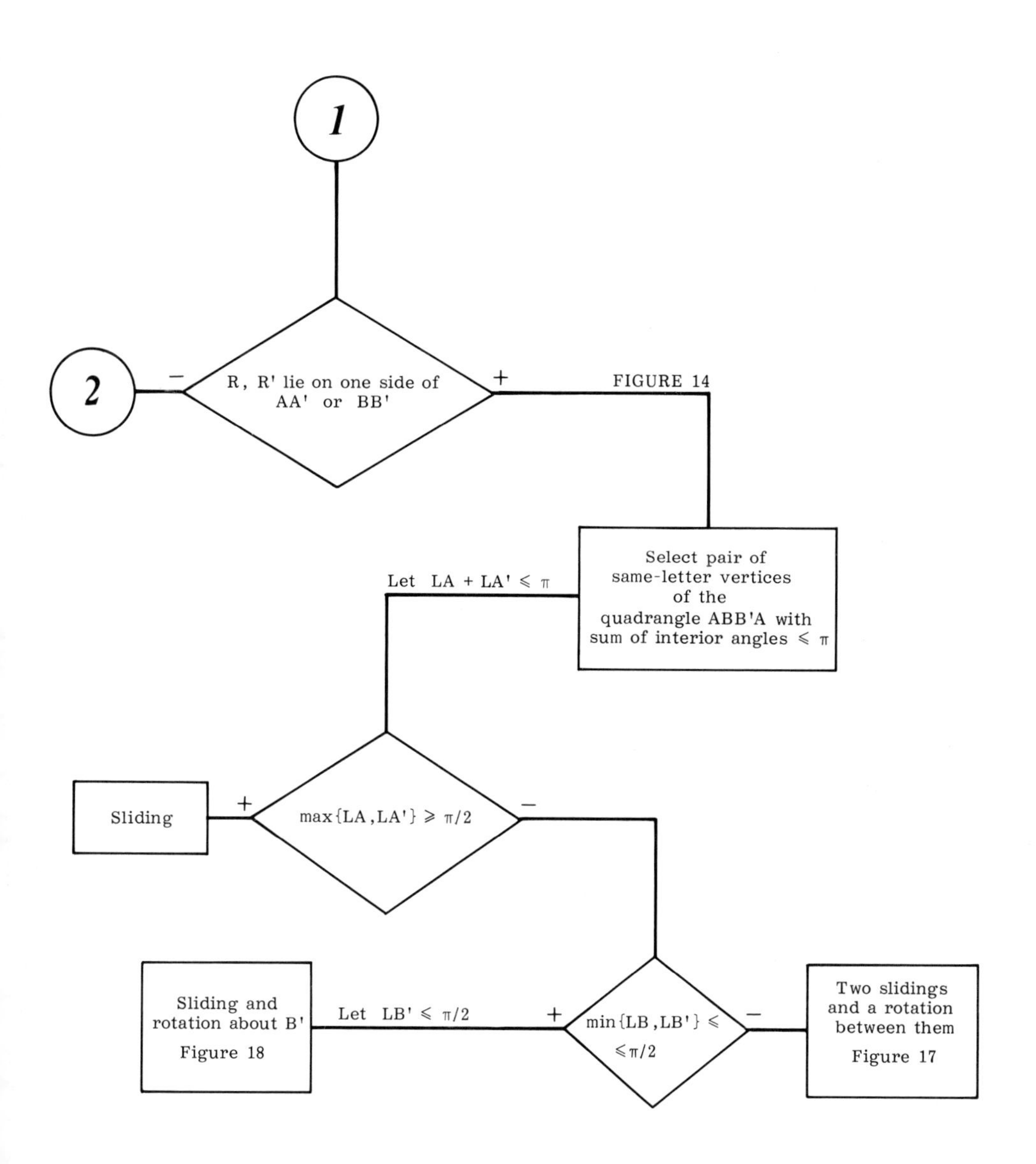
1
2
R, R' lie on one side of AA' or BB'
FIGURE 14
Select pair of same-letter vertices of the quadrangle ABB'A with sum of interior angles ⩽ π
Let LA + LA' ⩽ π
Sliding
max{LA,LA'} ⩾ π/2
Sliding and rotation about B'
Figure 18
Let LB' ⩽ π/2
min{LB,LB'} ⩽ ⩽π/2
Two slidings and a rotation between them
Figure 17

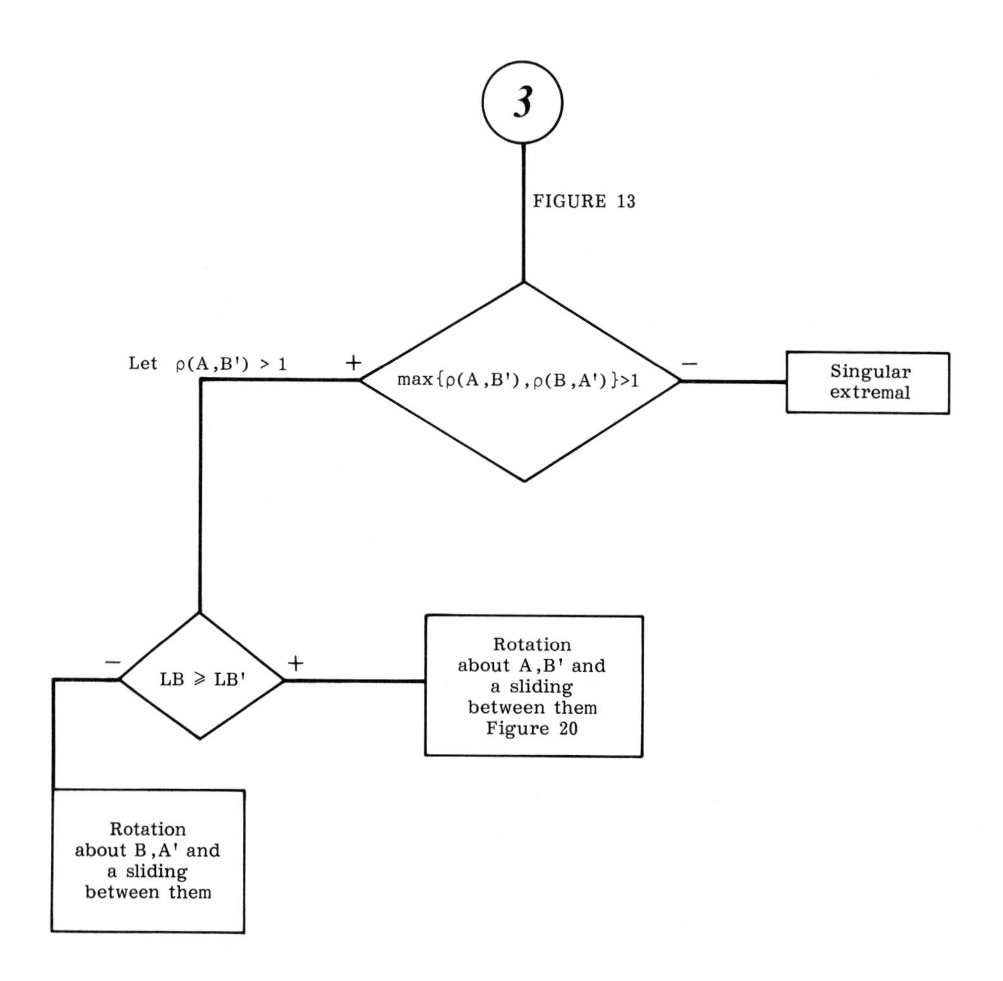

3
FIGURE 13
max{ρ(A,B'),ρ(B,A')}>1
+
−
Singular extremal
Let ρ(A,B') > 1
LB ⩾ LB'
−
+
Rotation about A,B' and a sliding between them Figure 20
Rotation about B,A' and a sliding between them

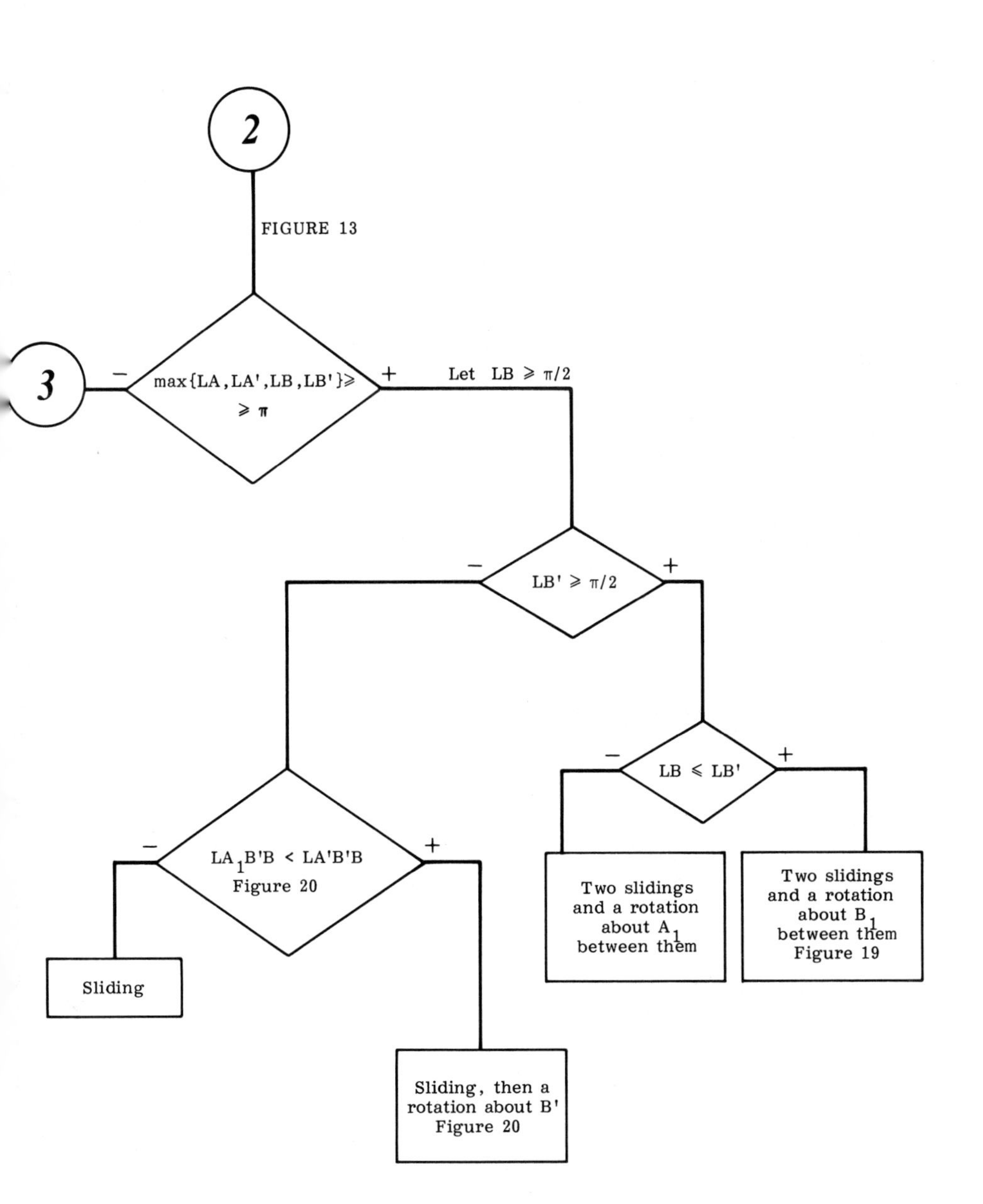
2
FIGURE 13
3
max{LA, LA', LB, LB'} ≥
≥ π
Let LB ≥ π/2
LB' ≥ π/2
LB ≤ LB'
LA1B'B < LA'B'B
Figure 20
Sliding
Sliding, then a
rotation about B'
Figure 20
Two slidings
and a rotation
about A1
between them
Two slidings
and a rotation
about B1
between them
Figure 19

APPENDIX

THE MAXIMUM PRINCIPLE IN OPTIMAL CONTROL PROBLEMS WITH REGULAR CONSTRAINTS

For the reader's convenience, we formulate the maximum principle for optimal control problems involving nonsmooth regular constraints on state and control. This maximum principle is the centerpiece of our solution of Ulam's problem. Nevertheless, at present it has not been treated in accessible literature. The proofs can be found in [7], [8].

We will consider the following general optimal control problem with unfixed time:

$$\begin{aligned} I(p) &\to \min , \\ \kappa_j(p) &\leq 0 , \\ K(p) &= 0 , \qquad (t \in [t_0, t_1], \quad i = 1, \ldots, d_G, \\ \dot{x} &= f(x,u,t) , \qquad j = 1, \ldots, d_\kappa) . \\ G_i(x,u,t) &\leq 0 , \\ g(x,u,t) &= 0 , \end{aligned} \tag{46}$$

Here $p = (x(t_0), t_0, x(t_1), t_1)$ is the terminal vector. The minimum in (46) is sought in the class of trajectories (x,u), where $x(\cdot)$ is an absolutely continuous function, and $u(\cdot)$ is a bounded measurable function. The given trajectory (x^0, u^0) is studied for a minimum. In the formulation (46) the functions I, κ_j, G_i are scalar functions, and x, u, K, f, g take values in finite-dimensional Euclidean spaces of dimension d_x, d_u, d_K, d_f, d_g, respectively.

Many optimal control problems, which prior to works [7], [8] were considered separately, reduce to the stated general problem (46). For example, in the optimal speed problem $I(p) = t_1 - t_0$. The optimal control problem with integrable minimizable functional

$$\int_{t_0}^{t_1} F(x,u,t)\,dt$$

reduces to (46) by means of introducing the additional state variable z, the additional differential equation $\dot{z} = F(x,u,t)$, and the function $I(p) = z_1 - z_0$.

Problem (46) is investigated under general assumptions pertaining solely to the nature of the functions describing the restrictions.

ASSUMPTIONS

- a. The functions f, g, K are continuously differentiable in their arguments;
- b. the function I, κ_j are locally convex in p;
- c. the functions $G_i(x,u,t)$ are continuously differentiable in x, t and locally convex in u;

• d. the constraints $G(x,u,t) \leq 0$, $g(x,u,t) = 0$ are regular in u for every point (x,u,t) belonging to the closure in measure of the trajectory (x^0,u^0), $t \in [t_0,t_1]$.

We will explain the terminology. We consider a continuous scalar function $\phi(y,z)$ where $y \in R^n$, $z \in R^m$.

DEFINITION 1. The function $\phi(y,z)$ is called locally convex in y if

• 1. for any y, z, $\bar{y}$ there exists a derivative in the direction $\bar{y}$ at the point y, z:

$$\phi'_y(y,z;\ \bar{y}) = \lim_{\varepsilon \to 0^+} \varepsilon^{-1}(\phi(y+\varepsilon\bar{y},\ z) - \phi(y,z)) \ ;$$

• 2. the function $\phi'_y(y,z;\ \bar{y})$ is convex in $\bar{y}$ and upper semi-continuous in y, z.

The smooth and convex functions belong to the class of locally convex functions. This class is closed under the operation max applied to a finite number of functions, and also under the operation of smooth change of variables.

DEFINITION 2. We say [8] that the point (x,u,t) belongs to the closure in measure of the trajectory (x^0,u^0), $t \in [t_0,t_1]$, if for any $\varepsilon > 0$ the inequality

$$\text{meas} \left\{\tau \in (t-\varepsilon,t+\varepsilon) \cap [t_0,t_1] \mid |x^0(\tau)-x| < \varepsilon,\ |u^0(\tau)-u| < \varepsilon\right\} > 0$$

is satisfied.

The union of all points of the closure in measure of the trajectory (x^0,u^0) forms a closed set. If the control $u^0(\cdot)$ is continuous, then this set coincides with the graph of the trajectory (x^0, u^0), $t \in [t_0, t_1]$.

DEFINITION 3. We say that the constraints $G(x,u,t) \leq 0$, $g(x,u,t) = 0$ are regular in u at the point (x^*, u^*, t^*) if there is a vector $\bar{u} \in R^{d_u}$ such that

$$G'_{iu}(x^*, u^*, t^*; \bar{u}) < 0 ,$$

$$g'_u(x^*, u^*, t^*)\bar{u} = 0 \qquad (G_i(x^*, u^*, t^*) = 0) .$$

In other words, regularity at (x^*, u^*, t^*) means that there exists a direction $\bar{u}$ reducing the active constraints $G_i(x,u,t) \leq 0$ that is a tangential direction to the constraint $g(x,u,t) = 0$.

A necessary condition of the first order for a strong minimum in (46) is the maximum principle.

THEOREM. Let (x^0, u^0) be a point of the strong minimum in (46). Then there exists a nontrivial set of Lagrange multipliers

$$\hat{\lambda} = (\alpha_I, \alpha, C, \psi, \psi_t, m, \lambda) ,$$

where

$$\alpha_I \in R_+ , \qquad \lambda \in L_\infty^{d_G}[t_0, t_1] ,$$

$$\alpha \in R_+^{d_K} , \qquad \psi \in W_{11}^{d_x}[t_0, t_1] ,$$

$$C \in R^{d_k} , \qquad \psi_t \in W_{11}^{1}[t_0, t_1] ,$$

$$m \in L_\infty^{d_g}[t_0, t_1] ,$$

and also a set of measurable functions

$$n_i \in L_\infty^{d_u}[t_0, t_1] , \qquad i = 1, \ldots, d_G$$

and vectors

$$\nu_I \in \partial I'_p(p^0; \cdot) \quad ,$$

$$\nu_j \in \partial \kappa'_{jp}(p^0; \cdot) \quad , \qquad j = 1,\ldots,d_\kappa$$

such that the following conditions are satisfied:

- 1. $\alpha\kappa(p^0) = 0, \quad \lambda \geq 0, \quad \lambda G(x^0,u^0,t) = 0,$

  $n_i(t) \in \partial G'_{iu}(x^0,u^0,t; \cdot);$

- 2. the local maximum principle:

$$f'^*_u\psi - g'^*_u m - \sum_i \lambda_i n_i = 0 \quad ;$$

- 3. the conjugate system:

$$-\dot{\psi} = \bar{H}'_x \quad , \qquad -\dot{\psi}_t = \bar{H}'_t \quad ;$$

- 4. the transversality conditions:

$$\psi(t_0) = \ell'_{x_0} \qquad \psi(t_1) = -\ell'_{x_1} \quad ,$$

$$\psi_t(t_0) = \ell'_{t_0} \qquad \psi_t(t_1) = -\ell'_{t_1} \quad ;$$

- 5. the maximum principle proper:

$$\psi(t)f(x^0,u^0,t) + \psi_t(t) = 0 \quad \text{for almost all} \quad t \in [t_0,t_1];$$

$$\psi(t)f(x^0,u,t) + \psi_t(t) \leq 0 \quad \text{for any} \quad u, \ t$$

satisfying the conditions $t \in [t_0,t_1]$, $G(x^0,u,t) \leq 0$, $g(x^0,u,t) = 0$.

All equalities involving t-measurable functions are understood to hold almost everywhere on $[t_0,t_1]$. In the formulation of the theorem,

$$\bar{H}(x,u,t,\psi,m,\lambda) \;=\; \psi f - \lambda G - mg$$

is the Hamiltonian;

$$\ell(p,\ \alpha_I,\ \alpha,\ C) \;=\; \alpha_I I + \alpha\kappa + cK$$

is the terminal Lagrange function.

We have denoted by ℓ'_{x_0}, ℓ'_{t_0}, ℓ'_{x_1}, ℓ'_{t_1} the components of the quasigradient ℓ'_p of the p-locally convex function ℓ with representation

$$\ell'_p \;=\; \alpha_I \nu_I \;+\; \sum_{j=1}^{d_\kappa} \alpha_j \nu_j \;+\; K'^{*}_p\, C \quad .$$

★

BIBLIOGRAPHY

[1] Ulam, Stanislaw M. *A Collection of Mathematical Problems*. New York: Interscience Publishers, 1960. (Russian translation: Nereshennye matematicheskie zadachi. Moscow: Nauka, 1964.)

[2] Rvachev, M.A. "On a Problem of the Ulam Type." *Funktsional'nyj analiz i ego primeneniya*, 11, 2 (1977): 58-66. (In Russian.)

[3] Rvachev, M.A. *On Minimizing an Integral of Displacement*. Avtoreferat dissertatsii. Moscow: MGU, 1974. (In Russian.)

[4] Gurevich, A.B. "The 'Most Economical' Displacement of a Segment." *Differential Equations*, 11, 12 (1976): 1583-1589. Published by Plenum Publishing Corporation. Translated from *Differentsialnye uravneniya*, 11, 12 (1975): 2143-2153.

[5] Penrose, R.K. "Extremal Motion of a Moved Line Segment." *Proceedings of the Montana Academy of Science*, 33 (1974): 46-51.

[6] Goldberg, Michael. "The Minimum Path and the Minimum Motion of a Moved Line Segment." *Math. Mag.*, 46, 1 (January 1973): 31-34.

[7] Dubovitskij, A.Ya., and A.A. Milyutin. *Necessary Conditions for a Weak Extremum in a General Optimal Control Problem.* Moscow: Nauka, 1971. (In Russian.)

[8] Dubovitskij, A. Ya. *Integral Maximum Principle in a General Optimal Control Problem.* Article deposited at VINITI. Moscow, 1974: 2639-2674. (In Russian.)

[9] Dubovitskij, A. Ya. "Solution of Ulam's Problem on Optimal Motion of Segments." *Izvestiya Akademii Nauk SSSR,* ser. matem., 40, 3 (1976): 673-684. (In Russian.)

INDEX

TRANSLATION SERIES IN MATHEMATICS AND ENGINEERING

A.A. Borovkov, Ed.
ADVANCES IN PROBABILITY THEORY: LIMIT THEOREMS FOR SUMS OF RANDOM VARIABLES
1985, approx 400 pp.
ISBN 0-911575-17-0 Optimization Software, Inc.
ISBN 0-387-96100-3 Springer-Verlag New York Berlin Heidelberg Tokyo
ISBN 3-540-96100-3 Springer-Verlag Berlin Heidelberg New York Tokyo

V.V. Ivanishchev, and A.D. Krasnoshchekov
CONTROL OF VARIABLE STRUCTURE NETWORKS
1985, approx 200 pp.
ISBN 0-911575-05-7 Optimization Software, Inc.
ISBN 0-387-90947-8 Springer-Verlag New York Berlin Heidelberg Tokyo
ISBN 3-540-90947-8 Springer-Verlag Berlin Heidelberg New York Tokyo

V.F. Dem'yanov, and L.V. Vasil'ev
NONDIFFERENTIABLE OPTIMIZATION
1985, approx 350 pp.
ISBN 0-911575-09-X Optimization Software, Inc.
ISBN 0-387-90951-6 Springer-Verlag New York Berlin Heidelberg Tokyo
ISBN 3-540-90951-6 Springer-Verlag Berlin Heidelberg New York Tokyo

A.N. Tikhonov, Ed.
PROBLEMS IN MODERN MATHEMATICAL PHYSICS AND COMPUTATIONAL MATHEMATICS
1985, approx 500 pp.
ISBN 0-911575-10-3 Optimization Software, Inc.
ISBN 0-387-90948-6 Springer-Verlag New York Berlin Heidelberg Tokyo
ISBN 3-540-90948-6 Springer-Verlag Berlin Heidelberg New York Tokyo

N.I. Nisevich, G.I. Marchuk, I.I. Zubikova, and I.B. Pogozhev
MATHEMATICAL MODELING OF VIRAL DISEASES
1985, approx. 400 pp.
ISBN 0-911575-06-5 Optimization Software, Inc.
ISBN 0-387-90948-6 Springer-Verlag New York Berlin Heidelberg Tokyo
ISBN 0-387-90948-6 Springer-Verlag Berlin Heidelberg New York Tokyo

V.G. Lazarev, Ed.
PROCESSES AND SYSTEMS IN COMMUNICATION NETWORKS
1985, approx. 250 pp.
ISBN 0-911575-08-1 Optimization Software, Inc.
ISBN 0-387-90950-8 Springer-Verlag New York Berlin Heidelberg Tokyo
ISBN 3-540-90950-8 Springer-Verlag Berlin Heidelberg New York Tokyo